American Institute of Chemical Engineers

Forest Products Division

Presentations at the
2007 AIChE Annual Meeting

November 4-9, 2007
Salt Lake City, Utah, USA

Printed from e-media with permission by:

Curran Associates, Inc.
57 Morehouse Lane
Red Hook, NY 12571
www.proceedings.com

ISBN: 978-1-60560-013-0

Some format issues inherent in the e-media version may also appear in this print version.

ISBN: 978-1-60560-013-0

American Institute of Chemical Engineers
Proceedings
Three Park Avenue
New York, NY 10016-5991
Phone: 212-591-8100

www.aiche.org

American Institute of Chemical Engineers

Forest Products Division

TABLE OF CONTENTS

154a High Performance Polymers Made with Renewable Resources

Joseph V. Kurian, E.I. DuPont de Nemours and Company, E.I. DuPont de Nemours and Company, Experimental Station, Wilmington, DE 19880

There is a growing need to develop renewably sourcedTM materials that provide performance and functionality equivalent to or better than petroleum based materials. DuPont is committed to research and development that will increase the use of renewable materials and reduce dependency on petroleum. There are many diverse applications for renewably sourcedTM products ranging from fabrics to carpets, inks and coatings, to construction materials and packaging. DuPont Engineering Polymers is moving forward with plans to produce a new family of high-performance thermoplastic resins and elastomer products made with renewable resources. The new products are DuPontTM Sorona® polymer and DuPontTM Hytrel® RS Thermoplastic Elastomers made with renewable resources. The key ingredient in Sorona® polymer is Bio-PDOTM, which is derived from corn sugar using a proprietary fermentation process. DuPontTM Hytrel® RS Thermoplastic Elastomer will be produced using a 100% bio-derived DuPont polyolTM, CerenolTM, made with Bio-PDOTM. DuPont/Tate & Lyle bioproducts LLC, a joint venture between DuPont and Tate & Lyle, started the first large-scale industrial production of Bio-PDOTM at the Loudon, TN facility in 2006. From a "cradle-to-gate" perspective, the production of Bio-PDOTM consumes 40% less energy and reduces green house gas emissions by 20% versus petroleum-based 1,3-propanediol. Production of 100 million pounds of Bio-PDOTM in the Loudon plant will save the equivalent of 10 million gallons of gasoline per year. The performance and processing characteristics of both Sorona® and Hytrel® made with renewable resources are as good as or better than those of current products made wholly from petrochemicals. This presentation will provide an overview of the Bio Based Materials Technology Platform, polymer properties, commercialization status and applications of these polymers. More information can be found at our web site: http://www.renewable.dupont.com.

154b Environmentally Friendly Polymeric Materials For Military Applications

John J. La Scala, Army Research Laboratory, AMSRD-ARL-WM-MC, Aberdeen Proving Ground, MD 21005

The Army has been moving towards the use of environmentally friendly materials for the past decade or so and is continuing to do so today. The Army determines and addresses environmental needs through the Army Environmental Requirements and Technology Assessments (AERTA). Due to recently enacted and proposed National Emissions Standards for Hazardous Air Pollutants (NESHAP), the Army has determined that the areas of coatings, adhesives, and composites are a high priority environmental need. As a result, the Army has been reformulating and evaluating new coatings, adhesives, and composite resin formulations to reduce volatile organic compounds (VOC) and hazardous air pollutant (HAP) emissions. Through the use of exempt solvents and new water-borne technologies, we have been able to formulate HAP-free and low VOC coatings formulations. Commercially available HAP-free adhesives are being evaluated to replace current HAP-containing adhesives in the military, including materials conforming to MMM-A-121 (rubber to metal bonding) and A-A-1936 (rubber cement). Successful HAP-free formulations will be demonstrated at Army depots for performance and effectiveness starting in 2008. The development of fatty acid monomers has allowed for the reduction of styrene HAP content in composite resins from ~45 wt% to 20-25 wt%. These resins are currently being demonstrated/validated for DoD use in composite replacement parts for Army tactical vehicles, including HMMWV hoods, Marines HMMWV helmet hardtop, an Air Force T-38 dorsal cover, and an MCM composite rudder for the Navy. Lastly, composite repair resins used for automotive and other repair applications, such as Bondo™, have been reformulated using fatty acid monomers to contain no HAP, while having similar performance relative to current materials.

KENAF FIBER REINFORCED BIOPOLYESTER COMPOSITES

S.Sahoo[1,], A.Nakai[1], M. Misra[2], A. K. Mohanty[3], Z. A. M. Ishak[4] & H. Hamada[1]*

[1]Advanced Fibro-Science, Kyoto Institute of Technology, Matsugasaki, Sakyo-ku, Kyoto606-8585, Japan

[2.] Composite materials and structures centre, Michigan State University, 2100 Engineering Building, East Lansing, MI, 48824-1226, USA

[3] School of Packaging, 130 Packaging Building, Michigan State Universities, East Lansing, MI 48824, USA

[4] School of Materials and Mineral Resources Engineering, Engineering Campus, University of Sains Malaysia, Penang, Malaysia

[*] Corresponding author:saswata.sahoo@gmail.com

ABSTRACT

Composites from biopolymer and biofiber becoming the value added materials from environment point of view. These materials have the potential to be the attractive alternative to the synthetic fiber composites considering both the cost and quality. In the present investigation composite specimens are prepared from biodegradable polymer matrix i.e. polyhydroxybutyrate-valerate (PHBV) and biofiber (kenaf) using injection molding technique. PHBV is the biodegradable biopolymer used as packaging materials. The objectives of the present work is to investigate the suitability of kenaf and bamboo fibers as reinforcing filler in PHBV composites, to study the effect of fiber content and effect of surface treatment (alkali treatment) of fibers on the mechanical properties and biodegradability of composites. Tensile properties of composites increased due to fiber loading and decreased due to alkali treatment (mercerization) of kenaf fiber. Impact strength of neat PHBV polymer was much higher than the composites. Impact strength of mercerized kenaf composites were higher than untreated fiber composites and increased with increase in fiber content. The tensile fractured surface morphologies of composite specimens were studied through SEM analysis. Change in mechanical properties, surface morphology and negligible degradation within ten weeks (70 days) of natural soil burial were observed from the investigation.

KEY WORDS: Kenaf, PHBV, composite, biofiber, SEM

INTRODUCTION

The manufacture, use and removal of traditional composite structures usually made up of synthetic fibers and non-degradable polymers are considered critically. Most commercially available plastics today are non-biodegradable and cause serious concern for the environment, associated with their disposal. Hence the use of renewable resources like biopolymers and biofibers (plant and animal-based) for the fabrication of biodegradable plastics is one of the many solutions to check plastic pollution. Biocomposites, prepared by reinforcing bio-fibers with environment friendly degradable biopolymers like polybutylene succinate (PBS), polylactic acid (PLA), cellulose acetate (CA), polyhydroxy butyl valerate (PHBV),polyester amide (BAK) etc. at low cost with acceptable mechanical properties are being developed which may be viable substitutes of the non-biodegradable plastic composites[1].Whole green composites or composites from biofiber and biodegradable polymer are attracting the attention of material scientists due to the increased environmental consciousness. The main constraint on the use of biodegradable polymers is the difference in the price of these polymers compared to bulk produced oil-based plastics. For scientists, the real challenge lies in finding applications which would consume large quantities of these materials to lead to price reduction allowing it to compete economically in the market.

There are some publications of biofiber based biodegradable polymer composites. Generally polylactic acid (PLA), polybutylene succinate (PBS), polyester amide (BAK), cellulose acetate (BIOCETA) etc were used for fabrication of composites. PHBV (trade name Biopol) has been used extensively in blending technology; however, its use in composite fabrication is just in an infant stage. PHBV (polyhydroxy butyrate co-valerate) is a co-polyester of poly (β-hydroxybutyrate) and 3-hydroxyvalerate. Few publications on PHBV composites have been reported [2, 3].

Natural fibers are renewable, abundant, eco-friendly, inexpensive and possess higher specific strength properties in comparison to synthetic fibers. Both biopolymers and biofibers can maintain a sustainable development of an economically and ecologically attractive structural composite technology. However the interfacial adhesion between the biofibers and biopolymers is not adequate because the biofibers are highly hydrophilic in nature, aggregate during processing and are not homogeneous like synthetic fibers[4]. They contain a lot of components like cellulose, hemicellulose, lignin, wax etc[5, 6]. Several workers have reported different types of surface modifications of the biofibers for imparting strength, homogeneity, degumming, moisture repellence, flame retardant properties etc. in order to enhance the interfacial adhesion between the components of the biocomposite[1, 6, 7].

In the present work, biofiber hemp has been used as the reinforcement with biopolymer PBS for the fabrication of biocomposite by using injection molding method.

EXPERIMENTAL

Materials:

PHBV: Biopol granules D411GN (Monsanto company, UK)
Kenaf: From Penang, Malaysia

NaOH: Nacalai Tesque, Kyoto, Japan

Chemical Modification:

Kenaf fibers were soaked with 5% alkali solution for 2 hours, and then washed with water having few drops of acetic acid followed by water.

Composite fabrication:

First polymer films were prepared by compression molding machine SHINTO (50T) at 165 ^{0}C. Then two layers of calculated quantities of fibers were kept between three polymer sheets and pressed again in the same compression molding machine at same temperature. Then the sheets so prepared are crushed and then compounded at 90-160 ^{0}C, 200 rpm screw speed by twin screw extruder JSW TEX 30 HSS. At last Kenaf-Biopol composites were prepared by ultra high speed injection molding machine PO YUEN, TM 30H, TOYO MACHINERY & METALS CO. LTD. (Germany) using these compounded pellets at temperature120-150 ^{0}C and injection pressure 50 kg/ cm^2 and injection speed 30mm/sec.

Testing:

Tensile properties of kenaf-PHBV composites were tested using Instron Universal testing machine 4206 (span length 115mm and speed 10 mm/min). Unnotched izode impact test (specimen length=60mm cut from original specimen) was done by Digital Impact Tester (TOYOSEIKI). SEM micrographs of tensile fractured surfaces were carried out through Scanning Electron Microscope JEOL JSM-5200. Soil burial degradation test of composites was carried out by burring the specimen prepared in soil under normal environmental condition.

RESULTS AND DISCUSSION

From Table 1 and Fig. 1 it is observed that tensile strength of neat polymer (control or 0 % fiber loading) is 25.23 MPa. With 10 wt% fiber (kenaf) loading tensile strength (TS) of composites decreased than control and increased there after for 15 wt% and 20 wt% fiber loading of untreated fiber. Similar trend was observed in case of alkali treated (mercerized) kenaf composites. It was observed that for same % fiber loading, tensile strength of alkali treated composites were lesser than that of untreated fiber composites. Almost no difference in mechanical properties was observed between untreated and alkali treated composites with 10wt % fiber loading. Tensile modulus of kenaf composites showed an increasing trend with increase in fiber content from 0 % to 20 % but tensile modulus decreased slightly with surface treatment (alkali treatment). This decrease of tensile properties with surface treatment may attribute to the decrease of individual fiber strength owing to chemical treatment during surface modification. Extension at break decreased with increase in fiber content and found higher value in case of alkali treated kenaf composites. It may attribute to the effective reinforcement and greater fiber matrix adhesion in case of higher % fiber loading and surface treatment. It is observed that impact strength of polymer is very high in neat polymer (22.51 KJ /M^{2}) and decreased drastically with fiber loading. In case of untreated fiber reinforcement, impact strength of composites showed a decreasing trend from 2.35 to 1.65 KJ /M^2 but an increasing trend was observed in case of alkali treated kenaf

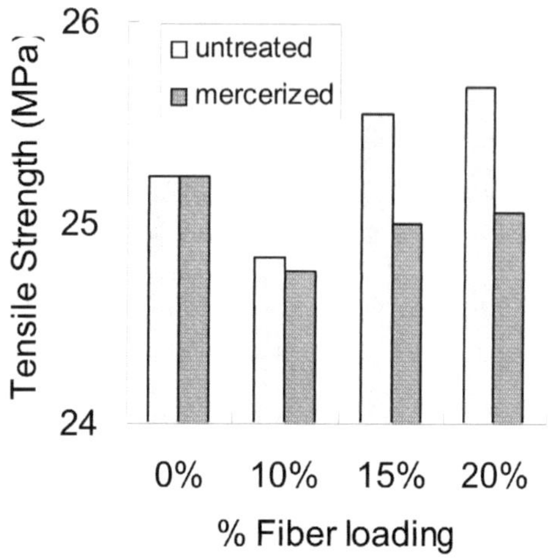

Figure1 Effect of fiber loadings and mercerization (alkali treatment) of fiber on the tensile strength of composites.

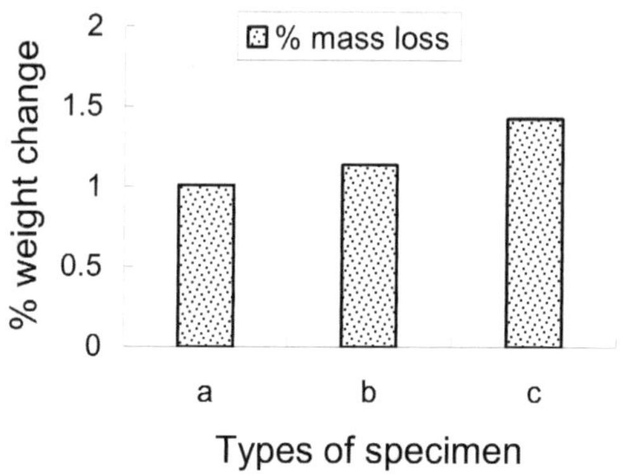

Figure 2 % Weight change of specimens (15% fiber loaded kenaf/pHBV composites after 70 days of burial (a= neat polymer b= natural, c= alkali treated)

Table 1 Mechanical Property of Kenaf-PHBV Composites

Kenaf fiber loading (wt %)	Tensile Strength (MPa)	Tensile Modulus (GPa)	Extension at break (mm)	Impact energy (KJ / M^2)
0 %	25.23	1.09	5.47	22.51
untreated				
10 %	24.83	1.9	2.35	7.0
15 %	25.54	2.22	1.91	6.42
20 %	25.68	2.42	1.65	5.97
Mercerized				
10 %	24.76	1.92	2.34	6.98
15 %	24.99	2.16	2.1	7.19
20 %	25.05	2.28	1.97	7.68

composites. Impact strength of alkali treated kenaf composites were found better than untreated kenaf composites due to increased fiber-matrix adhesion.

A very little weight change (1-1.5%) observed in the soill burial test of composites, as in Fig. 2, even after ten weeks (70 days). The specimens were in the soil during rainy season and recovered from soil during summer season. The % decrease of tensile strength is given in Fig. 3. The tensile strength reduction was 18.3, 15.9 and 19.2% respectively for neat polymer, 15% fiber loaded natural kenaf and 15% fiber loaded alkali treated kenaf composites respectively. The alkali treated kenaf composites degrade faster than natural and neat polymer composites. It is due to both polymer as well as fiber degradation in composites buried. It may be attributed to the more microbial attack to the alkali treated fiber composites owing to the removal of lignin, the cementing material of fiber, which makes the cellulose fiber more open so that microorganisms can recognize the cellulose chains easily.

Fig. 4a & b shows the SEM micrograph of 20 % fiber loaded composites. In case of 20% kenaf loaded composites, more number of holes due to fiber pullout was observed in untreated kenaf composites (Fig. 4a) indicating poor fiber-matrix adhesion. Lesser number of holes and fiber breakage is observed in Fig. 4b indicating a good fiber-matrix adhesion in case of alkali treated kenaf composites.

CONCLUSIONS

Biofiber-PHBV composites were prepared successfully using injection molding machine. Increase of kenaf fiber loading increased tensile properties and decreased extension at break. Impact strength of neat polymer was much higher than composites. Impact strength of composites decreased with untreated fiber loading and reverse trend was observed in case of alkali treated kenaf composites. Surface treatment decreased tensile strength and modulus and increased impact strength and extension at break as compared to untreated composites. Alkali treated kenaf composites showed more degradation in natural soil after 70 days of burial.

REFERENCES

1. A. K. Mohan, M. Mira and L. T. real., in Natural Fibers, Biopolymers and Biocomposites, Edited A. K. Mohanty, M. Misra and L. T. Drzal. CRC press, Taylor and Francis Group, FL, USA (**2005**) pp1-36
2. A. K. Mohanty, M. A. Khan, S. Sahoo and G. Hinrichsen, J. Mater. Sci., **35**, 2589 (2000)
3. S. Luo and A. N. Netravali, J. Mater. Sci., **34**, 3709(1999)
4. D. Nabi Saheb and J. P. Jog, Advanc. Polym. Technol. 18(4), 351 (**1999**)
5. M. Nakamura, K. Kitagawa, S. Sahoo, U. S. Ishiaku, A. Nakai, M. Kotaki and H. Hamada., Effects of interfacial adhesion on mechanical properties in injection molded

bamboo fiber/biodegradable polymer composites. Proceedings of the 4th International workshop on Green Composites (IWGC-4), (**2006**) September14-15, Tokyo, Japan.

6. A. Bismarck, S. Mishra and Thomas Lampke, in Natural Fibers, Biopolymers and Biocomposites, Edited A. K. Mohanty, M. Misra and L. T. Drzal. CRC press, Taylor and Francis Group, FL, USA (**2005**) pp 37-108
7. A. K. Mohanty, M. Misra, L. T. Drzal., Surface modification of natural fibers and performance of the resulting biocomposites: An overview. Compos. Interfaces. 8, 313 **(2001)**

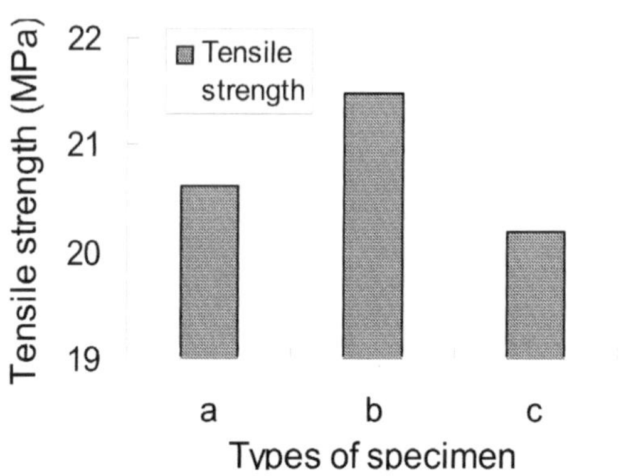

Figure 3 Tensile strength of specimens (15% fiber loaded kenaf/pHBV composites after 70 days of burial (a= neat polymer b= natural, c= alkali treated)

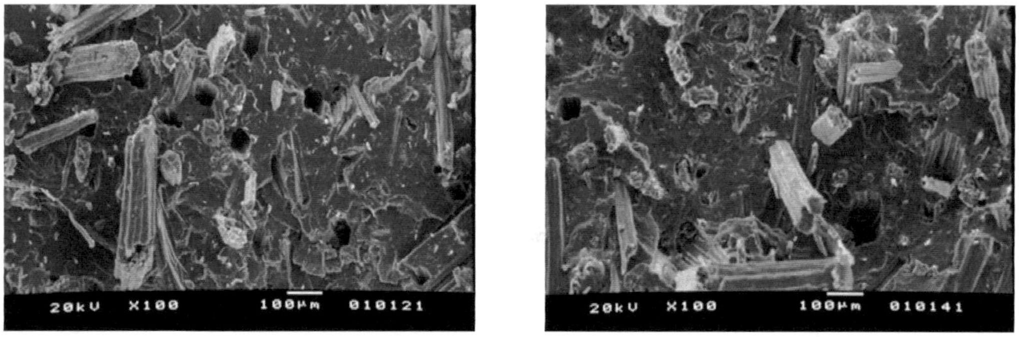

4a 4b

Figure 4 SEM micrograph of tensile fractured surface of 20 % fiber (kenaf) loaded composites, a) untreated b) mercerized

154d Starch-Oil Interaction In Dry Film Lubricants With Chemically Modified starch1

Girma Biresaw, USDA, 1815 N University street, Peoria, IL 61604 and R. Shogren, Plant Polymer Research Unit, National Center for Agricultural Utilization Research, Agricultural Research Service, United States Department of Agriculture, 1815 N. University Street, Peoria, IL 61604.

Starch is one of the most abundant farm-based raw materials. It is a significant component of such high volume commodity crops as corn, potato, rice, wheat, and barley. Because of the large surplus of these crops over demand, there is a great deal of interest in developing new uses for starch-based products. For some applications such as in lubrication, it is important that the starch dissolve in an appropriate solvent, preferably water. Unfortunately, starch is insoluble in water at room temperature and various methods are employed to improve its water solubility. One method involves conversion of free hydroxyl groups on starch glucose units into esters by reacting it with various anhydrides. Such chemical modification reduces the intramolecular H-bonding responsible for the high crystalinity, and, hence low solubility of starch. In the work described here, several chemically modified starches were prepared, formulated into dry film lubricants, and its friction properties investigated. The data was then analyzed to elucidate the mechanism of starch-oil interaction in such dry film lubricants.

1 Names are necessary to report factually on available data; however, the USDA neither guarantees nor warrants the standard of the product, and the use of the name by the USDA implies no approval of the product to the exclusion of others that may also be suitable.

154e Injection Molded Solid And Microcellular Polylactide-Carbon Nanotube Nanocomposites

Shaoqin (Sarah) Gong[1], Srikanth Pilla[1], Adam Kramschuster[2], Alexander Chandra[2], and Lih-Sheng Turng[2]. (1) Mechanical Engineering, University of Wisconsin-Milwaukee, 3200 North Cramer Street, Milwaukee, WI 53211, (2) Department of Mechanical Engineering, University of Wisconsin-Madison, 1513 University Avenue, Madison, WI 53706-1572

Poly(lactide) (PLA)-multiwalled carbon nanotube (MWCNT) nanocomposites were melt-compounded using a twin-screw extruder. Solid and microcellular tensile bar specimens were produced via conventional and microcellular injection molding processes, respectively. Various characterization techniques were applied to study the static and dynamic mechanical properties, degree of MWCNT dispersion, cell morphology, and crystallization behavior. The addition of a small amount of MWCNTs led to a decrease in the cell size and an increase in the cell density of the microcellular PLA. A transmission electron microscopy analysis of the PLA-MWCNT specimens revealed a higher degree of MWCNT dispersion in the microcellular PLA-MWCNT composite compared with its solid counterpart indicating the microcellular injection molding process further dispersed the MWCNTs. While the addition of 1.5 wt% MWCNTs slightly increased the average specific modulus of the microcellular PLA, it did not affect the average specific modulus of the solid PLA. In addition, for both solid and microcellular specimens, the specific strength, specific toughness and strain-at-break decreased with the addition of MWCNTs. The storage modulus was not affected significantly with the addition of MWCNTs, but was found to be higher for the microcellular specimens compared with their solid counterparts. Finally, the crystallinity of PLA increased significantly with the addition of MWCNTs.

Polyurethane Foams Using Functionalized Soybean Oil for Automotive Applications

Cynthia M. Flanigan, Christine Perry, and Deborah F. Mielewski.

Materials & Nanotechnology Department, Ford Motor Company, Ford Research and Advanced Engineering Laboratory, Dearborn, MI 48121

Renewable resources are becoming a promising alternative to petroleum products for automotive applications. These materials offer a reduced environmental footprint as well as advantages in cost savings and performance for select applications. Our work focuses on using functionalized soybean oil to produce flexible polyurethane foams. Flexible polyurethane foams are used extensively in the automotive industry for applications such as seating, instrument panels and headliners. Formulations and processing methods for soy-based foams were optimized for mechanical properties and odor. One of the main objectives of this study was to maximize the amount of soy content within the foam formulation while maintaining performance properties. Molded soy foam properties were compared to key properties required for interior automotive applications and showed promising results. We will review the research and development process of moving the technology from inception to implementation, with a discussion on formulation development, new synthesis methods to produce soy polyols, processing methods and physical properties of soy foam. In addition, the benefits of soy-based plastics, and technical and commercialization challenges will be discussed.

INTRODUCTION

Flexible polyurethane foam is used in many automotive applications including seat, headrest, headliner and instrument panel foam. In a typical North American vehicle, automotive manufacturers use approximately thirty pounds of foam in the interior cabin. With increasing interest to improve the environmental impact of automotive materials, efforts to utilize sustainable materials have drawn much attention. Currently, the materials used to form flexible polyurethane foams are derived from petroleum based chemical resources such as glycerin and ethylene oxide. The limited reserve and price fluctuations of fossil fuels and oils has prompted many companies to begin investigating renewable or agriculturally derived materials for their applications.

Soybeans are one promising raw material source for a variety of plastics due to their low cost and availability within North America. According to the USDA's statistics, in 2006, over 74.6 acres of soybeans were harvested, resulting in 23.2 billion pounds of soy oil. Soy oil pricing has been stable, ranging between 27.5 to 29.6 cents per pound for the past decade, making it attractive from an economic perspective[1]. Soybeans are comprised of about 18% oil, which can be chemically modified to make polyols for polyurethane foams.

In this project, we have focused our work on determining whether soy-based polyols can be utilized in the manufacture of flexible foams for interior applications. One of the inherent concerns of using these natural materials is the detectable odor. In our study, we investigate a novel method for manufacturing soy polyol, discuss odor reduction methods and evaluate the mechanical performance of molded foams.

FOAM FORMULATIONS AND MOLDING

In order to assess the properties and processing requirements of molding flexible foams that include soy polyols, foam formulations were developed within our laboratory and in partnership with Bayer

Corporation. Unlike rigid foams, flexible polyurethane foams are extremely sensitive to the quantity and nature of the formulation reactants and additives. Typical formulations include approximately ten different components. We have studied a commercially available MDI (diphenyl methane diisocyanate) seat cushion foam formulation to optimize the mechanical performance when soy-based polyol is substituted at various levels. Optimized formulations included 40 percent soy polyol in the polyol blend. Foams were produced by reacting polyol blends of petrochemical and soy based material with petroleum-derived diphenyl methane diisocyanate with appropriate combinations of blowing agent, amine/tin catalysts, cross-linker and silicone surfactant, as shown in Figure 1. The side reaction of isocyanate and water yields a urea linkage and evolution of carbon dioxide, which acts as the principal blowing agent in the foaming process. The targeted density of the foams was between 32-40 kg/m^3.

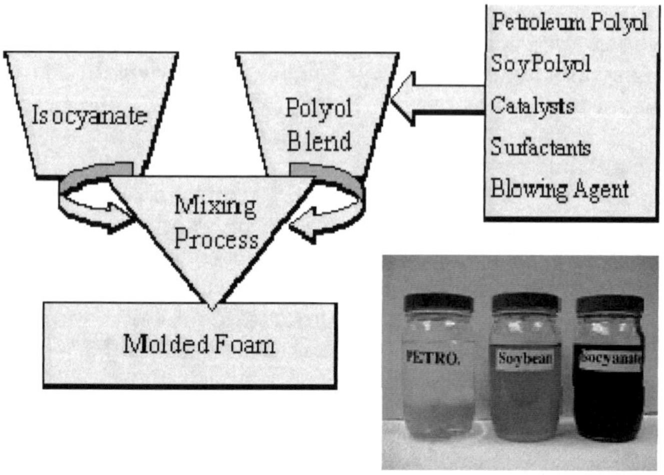

Figure 1: Process for molding soy-based polyurethane foam.

While the details of the formulation are proprietary, the chemicals for the foams included soy polyol, R2-052B from Urethane Soy Systems Company, polyether polyols and diphenyl methane diisocyanate from Bayer Corporation, amine catalysts and surfactants from Air Products and Chemicals, Inc. and GE Silicones. Chemicals were mixed in ratios according to Table I.

Table I: Basic Formulation for Flexible Soy Foam.

Components with Parts by Weight	
Soybean Oil Polyol	40
Petroleum Polyol	60
Water	3-4
Silicone Surfactant(s), Amine Catalyst(s), Cross-Linker(s)	1-3
Isocyanate	100 Index

Polyol blends, catalysts, surfactants and water were mixed together in a variety of formulation recipes. The components were blended with a 2300 rpm stirring head mixer before adding the MDI. The formulation was stirred until a color change in the viscous blend was noted, indicating the onset of creaming. Molded foams were poured into a heated aluminum box of 38.1 cm x 38.1 cm x 10.16 cm and sealed with a lid and clamps. Foams were cured at 70°C for 10 minutes, followed by room temperature cure for 24 hours before sectioning for testing.

NOVEL, SYNTHETIC METHOD TO FUNCTIONALIZE SOY OIL

Petroleum polyurethane foams are formed by reacting diisocyanate with a polyol blend, which includes a polyether polyol, surfactants, catalysts and blowing agent. Recently, several companies and universities have developed polyol products that are derived from crushed soybean oil instead of from glycerin or propylene oxide [2-4]. Soybean oil is a triglyceride with a fatty acid distribution of about 53 percent of linoleic acid (18 carbons and two double bonds) and 23 percent oleic acid (18 carbons and one double bond.) One inexpensive method to hydroxylate the soybean oil includes blowing air through the oil at elevated temperatures. With this method, undesirable side products are formed including aldehydes and ketones, which can contribute to a rancid odor in the polyol. For the blown soybean oil process, the hydrogens next to the double bonds within the soybean oil are removed, yielding free radicals. Oxygen from the air reacts at these radical sites on the lipid molecules to form peroxides. When heat is applied, the peroxides break down to form the hydroxyl groups and secondary oxidation products such as aldehydes. These low molecular weight products contribute to the undesirable odor and must be stripped out during a post processing step. Typically, this type of process requires either a metal catalyst, which is expensive, or temperatures greater than 82 °C to initiate the reaction. In our work, we have developed a novel method to synthesize soy polyols at room temperature, thus reducing the likelihood of unfavorable, secondary oxidative products.

A novel route we have taken has been to synthesize low odor, soy polyol from degummed cooking oil using a well-controlled reaction with ultraviolet light. Soy-based polyols were produced within our facility using degummed soy oil from Thumbseed Oil Producers. As shown in Figure 2, the soy oil is placed in a quartz column and is exposed to ultraviolet light within the range of 290 nm to 350 nm, as air is bubbled through the material at room temperature. This route enables us to add hydroxyl groups to the oil using a well-controlled reaction in which heat and metal catalysts are not needed.

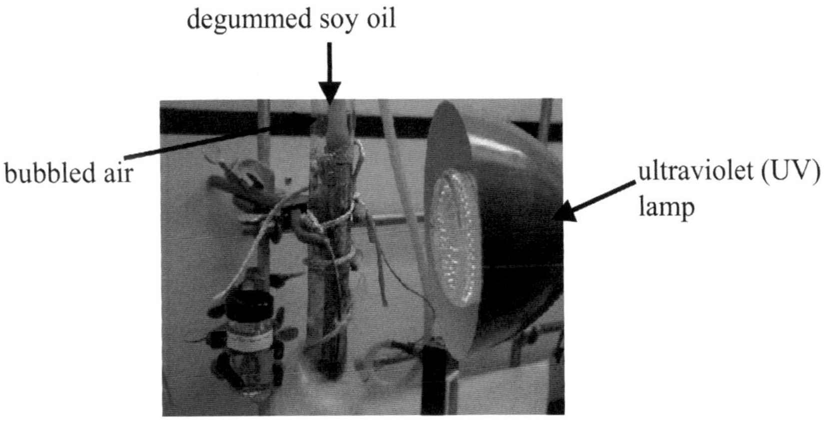

Figure 2: Laboratory setup to add hydroxyl groups to degummed soy oil using ultraviolet light.

The viscosities of the polyols, which reflects the degree of hydroxylation, were recorded as a function of reaction time and ultraviolet light exposure, seen in Figure 3. The data shows that for soy oil that is not exposed to ultraviolet light but has air bubbling through the column, there is no change in the viscosity of the material. However, with the introduction of ultraviolet light, there is a marked increase in viscosity around 180 hours, which correlates to the addition of hydroxyl groups in the oil. The change in hydroxyl groups was determined through standard titration methods. Overall, this method produced a lower odor polyol, as confirmed through odor test panel results in comparison to commercial soy polyol products with similar hydroxyl numbers.

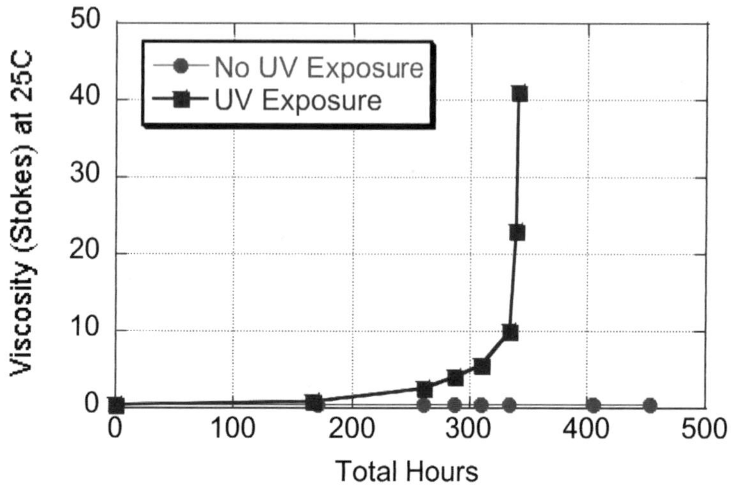

Figure 3: Measurement of viscosity of degummed soybean oil as a function of exposure to ultraviolet (UV) light.

A chemical analysis of the UV catalyzed soy polyol was completed, as shown in Table II. The samples showed a hydroxyl number in the range of 40-50 mg KOH/g, and viscosity of 2700 cps, which is suitable for molding flexible polyurethane foams. Also, the soy polyol exhibited excellent color characteristics (clear to slightly pale yellow color). The chemical results show that the UV method for catalyzing the hydroxylation of the soy oil produces an acceptable polyol for flexible foams.

Table II: Chemical analysis of UV catalyzed soy polyol.

Analysis	
Hydroxyl Number (mg KOH/g)	47
Peroxide Value (meq/kg)	202.3
Iodine Value (g/100g)	50.4
Acid Value (mgKOH/g)	5.4
Karl Fischer Water (%)	0.064

Our work has demonstrated a new method to hydroxylate soybean oil, which produces a low odor polyol that can be used for urethane foams. Since limited quantities of the UV catalyzed soybean oil were produced, the rest of the paper will focus on foams that were produced using commercially available soy polyols.

PROCESSING AND PHYSICAL PROPERTIES

After completing several formulation iterations using 5 to 40 percent soy polyol, soy-based foams were tested for physical properties and optical evaluation to optimize the foam. The trials and testing were completed as part of the technical development with Lear Corporation and Bayer Corporation. We have identified key parameters and processing conditions for scaling up from laboratory bench formulations to impingement mixed foams in production molds. Namely, catalyst and surfactant levels need to be adjusted to accommodate the efficient mixing of the production equipment and skinning effect on the foam surface. The physical properties of the foam were evaluated for each set of optimized formulations. Testing included density, indentation force deflection, tensile properties, elongation, tear resistance and compression set properties. Figure 4 depicts the processing steps to mold an automotive headrest part.

a) b) c)

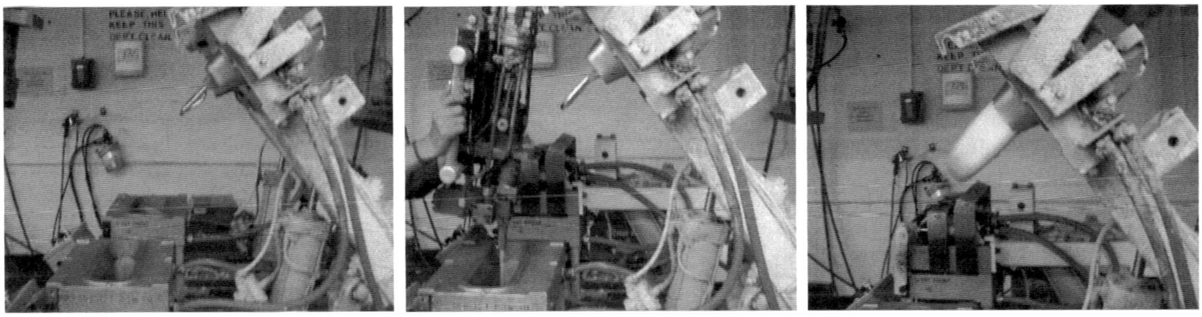

Figure 4: Process of forming headrest foam using an impingement mixing system. a) insertion of metal frame into tool lid b) pouring of chemicals into open mold c) demolding reacted polyurethane foam.

Test Methods:

Microscopy:
The optical appearance of the cellular structure of the foams was recorded using a LCD camera connected to a Nikon SMZ-U microscope. Images were taken at 50x magnification.

Density Measurements:
Foam blocks with dimensions 1"x2"x2" were cut from the center of the foam using a standard band saw. Samples were weighed and measured with a weighted foot dial gauge to calculate the density in kg/m^3. Four blocks per formulation were used in the density measurements.

Tensile Strength and % Elongation:
Following ASTM D3574 test procedures, specimens with 1" grip width and 5.5" total length were stamped from 12.5mm thick slabs using a tensile bar die. An Instron Model 5565 with 500N load cell in a tensile geometry was used to pull the samples at a cross-head velocity of 50mm/min. Tensile strength and percent elongation were recorded for five samples per set.

Compression Set for Constant Deflection:
Compression set values of the soy foams were determined following the ASTM D3574 standard. Foam samples were cut from the center of the molded foam into 1"x2"x2" blocks and measured with a weighted foot dial gage. Four samples per trial were compressed between 2 parallel plates to 50 percent deflection and heated in an oven for 22 hours at 70°C. After heat conditioning, dimensional measurements were taken again and compression set was calculated using the following equation:

$$C_d = \{(t_o - t_f)/\ t_o\} \times 100 \qquad \text{Equation 1.}$$

where C_d is the compression set at constant deflection, t_o is the original thickness and t_f is the final thickness of the sample.

Odor Test:
For interior automotive plastics, one key criteria that is used to assess odor is the SAE J1351 method. Samples were cut into 5.08 cm by 2.54 cm by 12.7 cm blocks and conditioned in a sealed, glass jar for one hour at 57°C. Five panelists evaluated and ranked the perceived odor of the samples on a scale from 1 to 5, according to the chart shown in Table III. A rating of "2" or lower is considered acceptable for these applications.

Table III: Odor Rating for Interior Plastics

Odor Rating Criteria SAE J1351	
1	No Noticeable Odor
2	Slight, but Noticeable Odor
3	Definite odor, but Not Strong Enough to be Offensive
4	Strong Offensive Odor
5	Very Strong Offensive Odor

Staining:

SAE J1885 test method was used to determine if soy-based foam contributed to staining on white vinyl test samples.

Fogging:
Following SAE J1756, automotive fogging tests assess the amount of volatile organic compounds that are released from materials at elevated temperatures. In the test, a block of foam is placed in a glass jar with a plate sealed with a rubber ring. The jar is heated and volatiles are collected on the top plate surface, which are then analyzed.

Physical and Optical Test Results

Optical micrographs provide a quick method to determine if the polyurethane foam has an open or closed cellular microstructure. Figure 5 shows a comparison of the microstructures in the petroleum and soy foams. The images indicate that the microstructure looks very similar between the two samples, and both foams exhibit an open cell structure, which is desirable for seating foams.

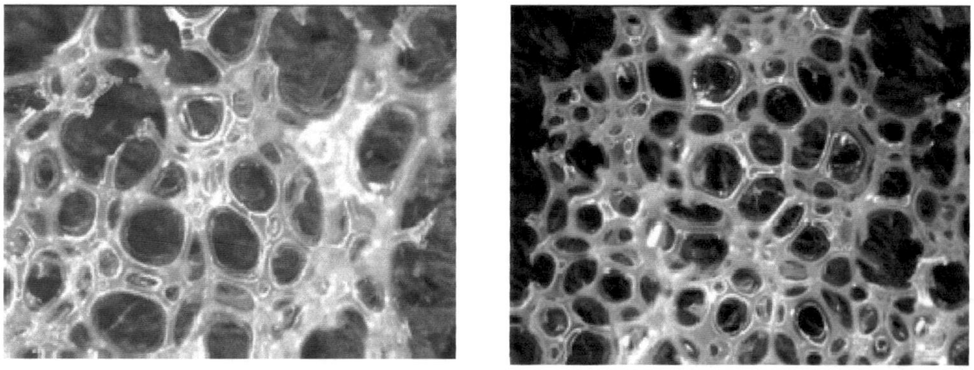

Figure 5: Optical micrographs of polyurethane foam cellular structure for petroleum (L) and soy (R) foam at 50x magnification.

In addition to the part surface appearance, the material properties of the soy foam were also evaluated. Several of the key screening properties include: detectable odor, staining of vinyl in contact with the foam, fogging of low molecular weight species and compression set. The properties of soy-based foam containing 27 percent soy polyol are compared to typical seating requirements in Table IV. The soy foam exhibited comparable properties to the benchmark standard in all of these categories.

Table IV: Physical Properties of Soy-Based Foam

Property Description	**Units**	**Target**	**Soy-Based Foam**	**Comments**
Density	Kg/m^3	35-40	37	**Pass**
Tensile Strength	kPa	82 min	174	**Pass**
Elongation	%	80 min	112	**Pass**
Compression Set	%	10 max (cushions)	2.8	**Pass**
Staining	Visual	No Stain	No Staining	**Pass**
Fogging	%	70 min.	98	**Pass**
Odor	Rating	2 max	2	**Pass**
Aged Flexibility	Visual	No Splitting or Crumbling after Aging	OK	**Pass**

Our results have shown that we have been successful in transferring the soy foam technology from laboratory molding equipment to production-type molding processes. In our project, we have overcome technical issues such as odor and part surface appearance in order to produce foams that pass the stringent automotive requirements.

SUMMARY

Soy based foams offer a sustainable alternative to petroleum based polyurethane foams. Not only do they utilize a renewable resource as one of its basic sources, but they offer the potential of being favorable from an economic perspective as well. Our results have demonstrated that these flexible, polyurethane foams containing soy may be formulated to pass the material requirements necessary for interior, automotive applications.

Once reformulated and optimized, flexible polyurethane foams produced from functionalized soybean oil exhibited very similar mechanical properties to those of the petroleum-based foams. Changing the amount and ratio of other chemicals used in the reaction is necessary to regulate the cell structure. As is typical of flexible, polyurethane foams, the cellular structure of the soy-based foam is highly dependent upon the foaming package. This package, which includes surfactants, catalysts, cell regulators, blowing agents and chain extenders, may be adjusted to improve the physical properties and appearance of the soy-based foam. Amine catalyst levels were increased from the current foam formulations to account for the less reactive nature of the soy material.

One of the inherent issues of using blown soy oil is that side products such as aldehydes and ketones can cause an undesirable odor in the functionalized soybean oil. In our work, we have identified a few key methods to reduce or eliminate this odor in the final product, including developing a new method to functionalize soy oil with low heat. By catalyzing the soy oil with ultraviolet light, we are able to control the hydroxyl number of the soy polyol and create a new method to produce low odor polyols from a renewable resource.

Implementing soy foam technology poses several technical challenges including the need to reformulate the foam for different interior applications such as seat cushions and headliner foams. In addition, to scale up the technology from laboratory scale equipment to production equipment, it is necessary to adjust the processing parameters and formulations to meet the current foam reactivity profiles. Blend stability of soy polyols and petroleum polyols may also be a concern depending upon the particular formulation and plant storage requirements.

Optimized foam formulations were able to meet benchmark standards for interior seating applications including compression set, tear resistance and tensile strength. Processing trials and physical properties showed that soy-based polyurethane foams are potential options for interior automotive applications.

REFERENCES

[1] http://www.ers.usda.gov/News/soybeancoverage.htm
[2] www.**soy**oyl.com
[3] http://news.dow.com/dow_news/prodbus/2006/20060925a.htm
[4] http://www.cargill.com/news/news_releases/2006/060306_urethanefoams.htm

ACKNOWLEDGMENTS

For this project, Brian Neal, Randy Carter and Don Schomer from Bayer Corporation and Asad Ali and Pratik Shah from Lear Corporation are thanked for all of their contributions. Additionally, we would like to thank the United Soybean Board, New Uses Committee for financial support.

217b Biodegradable Nanocomposites

Yulin Deng and Jihoon Lee. School of chemical and bio-molecular Engineering, Georgia Institute of Technology, 500, 10th st. N.W., Atlanta, GA 30332

Cellulose is the most important naturally abundant and renewable polymer. However, cellulose is a material that does not melt at the temperature lower than its degradation temperature. Strong intra- and intermolecular hydrogen bonds in the cellulose prevent its molecules from dissolution in most common solvents. For this reason, cellulosed nanocomposites have not been well studied. In this study, montmorillonite (MMT) nanoclay was used as the cellulose enhancement agent to improve the cellulose physical properties. N-Methylmorpholine-N-oxide monohydrate (NMMO) was used as the solvent to dissolve the natural cellulose.

The dispersibility of MMTs in pure cellulose was investigated using 7 different MMTs. Hydrophobicity of each MMT was compared and interaction force between MMT particles and cellulose matrix also was measured. The exfoliated and intercalated structure of the cellulose/MMT nanocomposites was observed by X-ray diffraction (XRD) and transmission electron microscopy (TEM). The tensile strength, modulus and thermal stability of nanocomposites increased with montmorillonite in the regenerated cellulose films. Changes of physical properties with different ratios of MMT contents were shown.

217c All-Cellulose Hierarchical Composites

Anthony Abbott, Julasak Juntaro, and **Alexander Bismarck**. Department of Chemical Engineering, Imperial College London, Polymer & Composite Engineering (PaCE) Group, South Kensington, London, United Kingdom

Considering the environmental and more stringent legislative pressures (at least in the EU), there is a pressing need for innovative, sustainable and recyclable materials. Therefore, the industry needs to move to greener materials that meet the physical performance of the conventional engineering materials. One of the most suitable and most abundant renewable materials able to fulfil this requirement is cellulose. Cellulose is a linear polysaccharide chain synthesized by plants or bacteria. We are aiming to produce composites, which combine cellulose from different origins in order to obtain an all-cellulose hierarchical composite structure. The different cellulose types used are commercially available microcrystalline cellulose powder, technical lignocellulosic fibres and bacterial cellulose. The bacterial cellulose was produced by bacterial synthesis using an Acetobacter Xylinum strain (BRP 2001) in highly ordered fashion, whereas, the plant cellulose is a semicrystalline polymer which also contains some other compounds like pectin, waxes and hemicellulose. The crystalline cellulose powder was dissolved in a non-derivative co-solvent (dimethylacetamide/LiCl) and subsequently regenerated to form the cellulose matrix. Natural (lignocellulosic) fibres were used as reinforcement. The natural fibre surfaces were modified by depositing nano-sized bacterial cellulose onto the fibres during the bacterial synthesis in presence of the fibres. This modification does not lead to any apparent mechanical degradation of the fibres. The attaching bacterial cellulose to the fibres enhances the effective fibre surface area of the fibres which is anticipated to lead to greater interaction with the regenerated cellulose matrix. The composite performance can be tailored by tuning the degree of crystallinity of the regenerated cellulose matrix, which is a function of the processing time during dissolution. It is anticipated that this type of composite materials can be produced using recycled paper as the cellulose source to produce the matrix of the composite. All-cellulose hierarchical composites might be able to substitute traditional based glass fibres reinforced polymers in a near future.

217d Coproduction of Wheat Straw-Derived Cellulose Fibers and Nanowhiskers with Ethanol in a Biorefinery

Christopher Saffron[1], Farzaneh Teymouri[1], Janette Moore[1], Darold McCalla[1], F. Larry Leistritz[2], and Bernie Steele[1]. (1) MBI International, 3900 Collins Rd., Lansing, MI 48910, (2) Agribusiness and Applied Economics, North Dakota State University, Morrill Hall 217, Fargo, ND 58105

Wheat straw is abundant in the Northern plain States of the U.S. such as North Dakota. Conversion of wheat straw to ethanol in a biorefinery could benefit State economies and provide transportation fuel from an inexpensive agricultural residue. Cellulosic fiber as a biorefinery coproduct could be used as a reinforcing agent in polymers. The value of the coproduct is a function of the physical property enhancement. Factors affecting process economics, such as saccharification yields, ethanol fermentation yields and cellulose fiber yields, will be discussed.

217e Biobased Materials from Polyhydroxyalkanoate and Thermoplastic Starch Blends and Their Biocomposites: Opportunities and Challenges

Amar K. Mohanty and Yashodhan S. Parulekar. School of Packaging, Michigan State University, 130 Packaging Building, East Lansing, MI 48824

Biobased as well as biodegradable materials were successfully developed through melt processing techniques from the blends of polyhydroxybutyrate-co-valerate (PHBV) and thermoplastic starch (TPS). The thermoplastic starch (TPS) was developed by blending of corn starch, glycerol, and a tough biodegradable polymer. The PHBV-TPS blends and their talc-filled biocomposites were successfully made into cast films. Further work explored the impact over time of three primary factors: ageing, leaching of the glycerol plasticizer, and the primary as well as the secondary crystallization of the PHBV component on the physical properties of the blends and their biocomposites. Mass transfer of moisture and oxygen through the developed films was correlated with the crystallinity and morphology of the materials using theoretical models. The developed materials were also exposed to aerobic composting conditions to study their biodegradable behaviors.

IMPROVED ISOLATION AND MELT PROCESSED PRODUCTS OF ZEIN[1]

Gordon W. Selling* and Kristen Kruger Woods
Plant Polymer Research Unit, National Center for Agricultural Utilization Research, USDA/Agricultural Research Service, 1815 N. University Street, Peoria, IL 61604
*gordon.selling@ars.usda.gov

ABSTRACT:

In order for zein, a potentially significant co-product of the bio-ethanol industry, to be used in new markets, improved isolation methods and improved zein based articles are needed. We have evaluated the ability of acetic acid to remove zein from corn gluten meal, distillers dried grains and ground corn. Acetic acid was found to remove zein more quickly, at lower temperatures and in higher yields when compared with alcoholic solvents. After 60 min at 25 °C, approximately 50% of the zein in corn gluten meal was removed. The protein composition of the zein removed from corn gluten meal using acetic acid is very similar to that of commercial zein by SDS-PAGE. Films and electrospun fibers could be produced directly from the zein acetic acid solutions generated from corn gluten meal extraction. To provide improved product properties, glyoxal, in the presence of base, has been used to cross-link zein in melt processes. Both batch and by reactive extrusion melt processes were performed giving articles with improved properties. The melt processed glyoxal modified zein require additional heating in order to deliver the best properties. Improved properties could be realized with glyoxal amounts as low as 1% and base amounts as low as 0.19 μmoles of NaOH per gram of zein. The resulting zein articles had improved tensile strength and reduced solubility.

INTRODUCTION

Zein is located in protein bodies distributed through the endosperm of corn (Lawton 2002). Zein is potentially a major co-product of the bio-ethanol industry. In a wet-milling operation it is found in corn gluten meal (CGM). In a dry milling operation it is found in distillers dried grains (DDG). There are four main classes of zein: α, β, γ and δ. The β, γ and δ zeins are found mainly in the wall of the protein body, while the α zein is located inside the protein body (Lawton 2002). Historically, zein was primarily used in the textile fibers under the trade name VICARA® as a wool alternative (Shukla and Cheryan 2001; Lawton 2002). Zein needed to be derivatized with formaldehyde to be suitable in the textile market (Swallen 1939;Croston et al 1945; Cline 1949; Evans and Croston 1949). Currently, zein is mainly used to coat tablets or food (Shukla 1992). Improved routes to deliver zein in a cost effective fashion are being developed (Shukla et al 2000; Cheryan 2001; Parris and Dickey 2001; Lawton 2006; Cheryan 2006A; Cheryan 2006B).

---

[1] Names are necessary to report factually on available data; however, the USDA neither guarantees nor warrants the standard of the product, and the use of the name by USDA implies no approval of the product to the exclusion of others that may also be suitable.

The current methods for isolating zein from corn gluten meal (CGM) involve extraction with either iso-propanol (IPA) or ethanol (EtOH) (Carter 1970; Shukla et al 2000; Cheryan 2001; Cheryan 2006A; Cheryan 2006B; Lawton 2006). While commercial operations may be developed using these techniques, depending on the downstream processing that may be utilized, costly zein isolation may be required. Recently zein has been modified using various electrophilic reagents that would require zein isolation for them to proceed in good yield (Wu et al 2003A; Wu et al 2003B; Biswas et al 2005A; Biswas et al 2005B; Biswas et al 2006; Sessa et al 2007). While there are many good solvents for zein (Lawton 2002), many of them have properties that make them less than ideal for zein isolation. Many of the solvents described have nucleophilic functionality that may react with the desired electrophile being added downstream. This would include extracting solvents such as ethanol or ethylene diamine. Other extracting solvents which may not be very reactive may have boiling points which are too high, making isolation of the final product costly. Solvents having this attribute would include N,N-dimethylformamide or N-methylpyrrolidone. Acetic acid (AcOH) is a good solvent for zein (Sessa et al 2003; Selling et al 2007; Sessa et al 2007), and has a relatively low boiling point (118 °C). AcOH has successfully been used to remove protein from kafirin (Taylor et al 2005), however properties of the resulting protein, such as film quality, were not determined. The nucleophilic functionality present in AcOH is relatively unreactive making it less nucleophilic than many of the solvents listed. This suggests that the resulting AcOH zein solution may be suitable in some of the chemistry currently detailed in the literature for zein derivatization, preventing the need for zein isolation.

As detailed previously, formaldehyde was the main reagent used for derivatizing zein so that it would be suitable in the textile fibers market. Commercial use of formaldehyde requires significant investment to be performed safely. Therefore, other chemistries are being explored as an alternative to formaldehyde to improve the properties of zein based materials (Yang and Wang 1996; Wu et al 2003A; Wu et al 2003B; Kim et al 2004; Lee at al 2005; Sessa et al 2007). However, these routes to modify the properties of zein based materials may utilize chemicals of a similar toxicity to formaldehyde, or they have not been shown to be resistant to dissolution into good zein solvents. Other proteins have also been derivatized with aldehydes, typically glutaraldehyde, such as peanut (Liu et al 2004), cotton seed (Marquie 2001), wheat (Hernandez-Munoz et al 2004; Hernandez-Munoz et al 2005), and soy (Wang et al 2007).

Of the potential crosslinking reagents that have been evaluated, glyoxal (GLY) may have the most promise of providing a crosslinked protein (Lee et al 2005). Previous research using GLY to crosslink proteinaceous materials have shown varied results (Paetau et al 1994; Marquie 2001; Lee et al 2005, Vaz et al 2005). The use of GLY to crosslink zein in solution has recently been described (Woods and Selling 2007). The use of GLY, or any crosslinking reagent, to cross-link zein in the melt state has not been described in the literature.

Developing new technologies that will provide a low-cost route to isolate zein and allow zein to enter new markets will improve the economics of the bio-ethanol industry and allow precious non-renewable feed stocks to be used in other markets. We report our efforts to determine if AcOH can extract zein from CGM more efficiently than incumbent processes. We will also be detailing our efforts around the use of GLY to provide an improved zein based article where the process is carried out in the melt state.

MATERIALS AND METHODS

Materials and Equipment

Commercial zein was obtained from Freeman Industries (Tuckahoe, NY), Lot F40004081 (14.1% nitrogen, 5.0% moisture, 0.57% ash) herein designated as Lot 4081. CGM was obtained from Aventine Renewable Energy, Inc. (Pekin IL). The % protein of the CGM was determined by multiplying the % nitrogen by 6.25, giving 71.7% after correcting for moisture. DDG was obtained from Big River Resources LLC (West Burlington, IA) and had 4.01% nitrogen which gives a protein content of 27.6%. Ground corn was obtained from Bunge Milling (Danville, IL) and had % nitrogen of 1.25 which gives a protein content of 8.7%. Other solvents and reagents were obtained from Sigma-Aldrich (St. Louis, MO). NuPAGE Novex 4-12% Bis-Tris gels, lithium dodecyl sulfate (LDS) sample buffer and protein molecular weight standards were obtained from Invitrogen (Carlsbad, CA). Infrared (IR) spectra were obtained on a Thermo Nicolet Avatar 370 FTIR where the samples were tested in KBr pellets. The electrospinning power supply was a Gamma Model ES50P-10W (Gamma High Voltage Research, Inc., Ormond Beach, FL). The syringe pump was a KDS 100 Series (KD Scientific, Holliston, MA) using 5 mL syringes with 1" long flat tipped 17 gauge needles. A quick connector, DR18-250-C (Panduit, Corp., Tinley Park, IL) was used to connect the hot wire from the power supply to the needle. The target was a 12" square sheet of 1/8" thick copper, covered in aluminum foil and held upright in a non-conductive material. Infrared spectra were obtained on a Thermo Nicolet Avatar 370 FTIR where the samples were tested in KBr pellets. Scanning electron microscopy (SEM) was performed after sputter coating the fibers with Au–Pd and then examined by a JEOL 6400 V (JEOL USA, Inc., Peabody, MA). A Haake Fisons Rheocord 90 torque rheometer using the 600 mixing bowl (Thermo Electron Corporation, Madison, WI), was used to make the torque measurements on the blends. Torque curves presented have been smoothed using one and one half minute moving averages. High shear roller rotors were used. The extruder was a Werner Pfleiderer ZSK30 corotating twin-screw extruder (Coperion, Ramsey, NJ) with 14 barrel sections (length/diameter ratio = 44), including the feed section. The screw configuration is described elsewhere (Willet and Finkenstadt 2003). For injection molding of the extrudate from the twin screw, an ACT75B injection molder (Cincinnati Milacron, Batavia OH) can be used with a suitable mold (Master Precision Mold, Greenville, MI) producing Type I tensile bars in accordance with ASTM D638-99.

Extraction of Zein from Corn Gluten Meal, Distillers Dried Grains or Ground Corn

To 25 grams of CGM, or other corn product, was added 75 grams of the desired solvent. This mixture was then allowed to stir at the desired temperature for the desired period of time. The mixture was then poured into centrifuge tubes and spun at 3000 rpm for 20 minutes using a Beckman GS-6R centrifuge (Fullerton, CA). A portion of this supernatant was then removed and allowed to dry overnight at room temperature, followed by additional drying at 105 °C for two hours to provide the % solids, and thus the % zein, present in the supernatant. Percent nitrogen data and IR spectra were obtained from the dried zein sample.

Gel Electrophoresis

Denaturing gel electrophoresis was conducted using zein obtained from the dried supernatant collected after the extractions. Samples for gel electrophoresis were dried without added heat. Zein was dissolved in DMF and then diluted to an appropriate concentration with water and loading buffer. DTT was added and the samples were boiled for 5 minutes. Samples were loaded onto a NuPAGE Novex 4-12% Bis-Tris gel and run at 100V for 40 minutes. Gels were stained with Coomassie Blue or silver.

Preparation of Zein Films and Fibers

Zein films from commercial zein were prepared by dissolving the desired amount of zein in AcOH and stirring for 30 minutes at room temperature. The solution was then cast using a 0.05 inch film knife. For zein produced using AcOH extraction, the AcOH supernatant was used to cast a film using a 0.05 inch knife. The films were allowed to dry for 3 days before cutting samples for tensile property testing.

For the production of electrospun fibers, the method previously described was used (Selling et al 2007). Briefly, the desired amount of zein solution was pumped at 12 mL/hour using a syringe pump. A 17 gauge 1 inch needle with a flat tip having a quick disconnect terminal attached to it was threaded onto the syringe. The hot wire from the power supply was attached to the quick disconnect terminal. The copper target was covered with a sheet of aluminum foil, and an alligator clip was attached to the target so that both the foil and the copper were in contact and then the ground wire was attached to the ground on the power supply. The distance from the target to the needle tip was 10 cm and the power supply was set to 25 kV.

Tensile property testing – Solution Cast Samples

Samples were cut from films using a die that would produce bars with dimensions in accordance with ASTM D-638-V Standard. Five bars from each film were then placed at 50% RH (23 °C) for 3 days. Bar thickness was measured with a micrometer (Model No. 49-63, Testing Machines Inc., Amityville, NY) at three places within the testing region and averaged. Testing was done using an Instron Universal Testing Machine (Model 4021, Canton, MA) to determine the tensile strength (TS), Young's Modulus (YM) and percent elongation at break (%E). The Instron crosshead speed was 10mm/min with a gauge length of 7.62mm.

Blend preparation and melt processing for producing cross-linked zein

Reagents were added to 45 grams of zein and the blend was initially stirred with a spatula to provide a crude blend. This blend could be added to the Haake rheometer in approximately thirty seconds. The start time (T = 0) for all tests is thirty seconds after the start of sample addition. GLY was evaluated as a cross-linking reagent in the range of 1 to 6% (w/w dry-basis zein). Sodium hydroxide (NaOH) solutions used in this study were in the range of 0.001 to 1.5 M. The amount of NaOH that was delivered to the blend was in the range from 0.00019 to 0.279 mmoles/gram zein. Potassium hydroxide (KOH) was used as a 1 M solution and was used at 0.186 mmoles/gram zein. Calcium hydroxide

($Ca(OH)_2$) was used as a 0.5 M solution and was used at 0.093 mmoles $Ca(OH)_2$ /gram zein. Initial temperature of the chamber walls of the torque rheometer was varied between 40 and 90 °C, and the rotors were run at 50 rpm. The melt was allowed to mix for between 2 and 10 minutes. The majority of samples were processed at 80 °C for 10 minutes. These conditions should be assumed unless noted otherwise.

Reaction extrusion and injection molding of cross-linked zein

To the feed throat of a Werner and Pfleiderer co-rotating ZSK30 twin screw extruder was added 71.25 gr/min of zein (dry basis). To this material through the first injection port was added 13.2 gr/min of 1 M NaOH in water. Through the second injection port was added 3.75 gr/min of triethylene glycol (TEG, which functioned as a plasticizer to improve extrusion processing). Through the third injection port was added 3.8 gr/min of 40% glyoxal in water. For the control, water was used in place of the 40% glyoxal in the amount of 6.4 gr/min. The extruder can be run at 147 rpm. The first zone of the extruder was heated to 37.8 °C with the remaining zones heated to 80 °C. The specific mechanical energy put into the resin was 250 J/gr. The extrudate was collected as it exited the extruder – no die was used. The extrudate was ground using a Thomas Wiley mill to provide an appropriately sized mixture for injection molding. The extrusion zones in the ACT75B injection molder zones were set to 82, 116, 132 and 138 °C. The maximum force of the mold was 186.16 MPa. The mold was set to 52 °C. The articles were removed from the mold and tested for resistance to dissolution in AcOH.

Percent loss/Swelling ratio testing

Compression molded physical property samples were tested for resistance to dissolution in acetic acid (AcOH) or other solvents by placing approximately 0.5 grams of a molded bar in 10 grams of solvent. The sample remained in the AcOH for 16 hours, then the swelled bar was removed and gently blotted dry to obtain the wet weight (Ww). Wet samples were placed in a fume hood for 8 hours and then dried in a 130 °C oven for 24 hours before weighing to obtain the dry weight (Wd). Percent loss was calculated by comparing the final dry weight to the original sample weight. Swelling ratio was calculated using Equation 1.

$$\text{Swelling ratio } (Q) = (Ww - Wd)/Wd \qquad (1)$$

A swelling ratio of 1.0 describes a sample that has swelled 100% (doubling its mass when wet) compared to the dry bar. It is therefore possible to have a swelling ratio of less than 1.0, describing a sample that has swelled, but at less than a 100% change.

Tensile property testing – Melt Processed Samples

For the production of compression molded samples, the dough produced from the Haake was frozen with liquid nitrogen and then ground to a powder. The mold could produce four test samples at a time. Sufficient powder (1.5 g) was placed into each mold to obtain a well formed sample. Molding conditions were selected to provide a tensile test sample that did not have visible remnants of the starting powder. The mold was heated at 127 °C and a pressure of 28 MPa for 20 minutes, and then allowed to cool at room temperature. Samples were stored at 23 °C and 50% relative humidity for five days before testing physical properties. Data values reported are the average of five tested bars.

RESULTS AND DISCUSSION

Zein Extraction

CGM was added to AcOH at room temperature and allowed to mix for various times followed by isolation of the supernatant. The supernatant was then analyzed to determine % solids and % nitrogen. As can be seen in Fig. 1, AcOH quickly removes zein from CGM. The amount of zein extracted versus time can be accurately fitted using an exponential relationship.

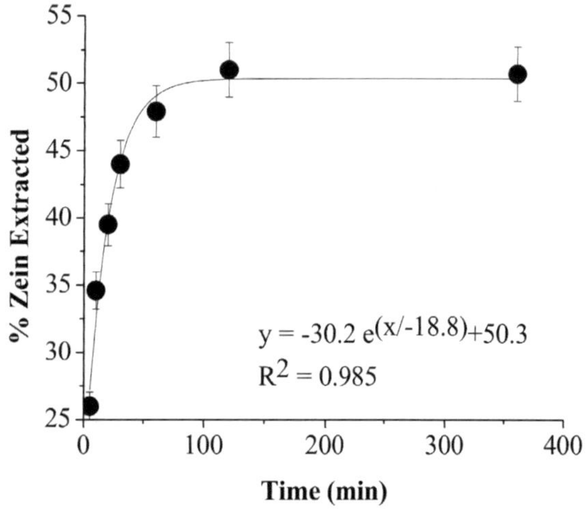

Fig. 1. Impact of time on % zein extracted from CGM using AcOH at room temperature and at 25% CGM solids.

The ability of AcOH to remove zein from CGM was compared with other solvents. AcOH is more effective at removing zein from CGM, giving 44% zein yield, versus 80% EtOH/water (3% zein yield), 90% EtOH/water (8% zein yield) or 60% IPA/water (32% zein yield) when the extraction was run at 25% solids, 25 °C and for 30 min. AcOH was also used to extract zein from DDG and GC, where 15 and 11% zein yields were obtained respectively. The % protein for the zein obtained from AcOH extraction was 84% (determined by multiplying the % nitrogen found by 6.25). A large amount of oil was also extracted from the DDG as well as evidenced by IR spectra (data not shown) and the fact that an oily grainy residue, instead of a solid, was isolated.

In order to determine the impact of temperature on zein extractability, the solvent temperature was increased from 25 to 70 °C at 25% solids and an extraction time of 30 minutes. There is little difference in the amount of zein that is extracted between 25 and 40 °C and between 55 and 70 °C, however there is an increase of 20% between 40 and 55 °C. Previously performed solution rheology studies have shown that significant structural differences take place above 45 °C (Selling et al 2005). Additionally it has been shown that with increasing temperature both the secondary structure (Cabra et al 2006; Selling et al 2007A) and tertiary structure (Selling et al 2007A) change. Therefore, the step increase in zein extracted from the samples may be due to changes in the structure of zein which will alter its rate of diffusion across the wall of the protein body.

In order to determine if there is a difference in the protein composition of the zein obtained from CGM using AcOH extraction, SDS-PAGE was performed using both Coomassie Blue and silver staining (Fig. 2). As can be seen, the protein composition of the zein obtained from CGM using AcOH is very similar to commercially produced zein as well as that obtained when using other extracting solvents. It can also be seen that the composition of the protein does not change significantly as the extraction time is increased from 30 to 120 minutes. IR analysis showed the presence of a shoulder in the carbonyl region (data not shown) which is indicative of the presence of fatty acids and oil, the presence of which would drive the % protein downward.

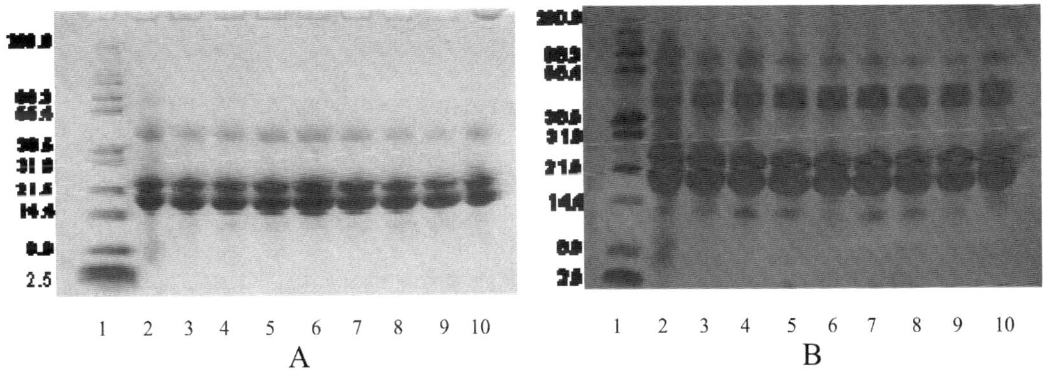

Fig. 2. SDS-PAGE gel of commercial zein and AcOH extracted zein from CGM. Gels stained with either Coomassie Blue (A) or silver (B). Lane: 1) M.W. standards; 2) Commercial zein; 3 & 4) AcOH 180 min extraction; 5 & 6) AcOH 30 min extraction 7) 60% IPA 30 min extraction; 8) 60% Acetone-water 30 min extraction; 9) 80% ethanol-water 30 minute extraction 10) 80% ethanol-water 180 minute extraction.

In order to determine if the zein produced via AcOH extraction of CGM can provide a product with useful properties, films were prepared from AcOH solutions of commercial zein and zein obtained from CGM using AcOH as the extracting solvent. The TS, %E and YM of films of commercial zein are 43 MPa, 13% and 777 MPa, respectively, when tested at 50% relative humidity. Zein isolated using AcOH provided films having TS, %E and YM of 18 MPa, 11% and 293 MPa, respectively, at 50% relative humidity. The presence of higher levels of fatty acids and esters would contribute to the reduction in tensile strength.

To determine if the AcOH solution obtained after CGM extraction could be used to produce fibers, the resulting AcOH solution from extracting CGM had its solids increased to 29% by evaporation and then it was immediately electrospun (Selling et al 2007B). The spinning performance of the AcOH extracted zein solution was equivalent to fresh AcOH solutions made using commercial zein, with both producing round birefringent fibers having diameters of approximately 0.4-1.6 μm (Fig. 3).

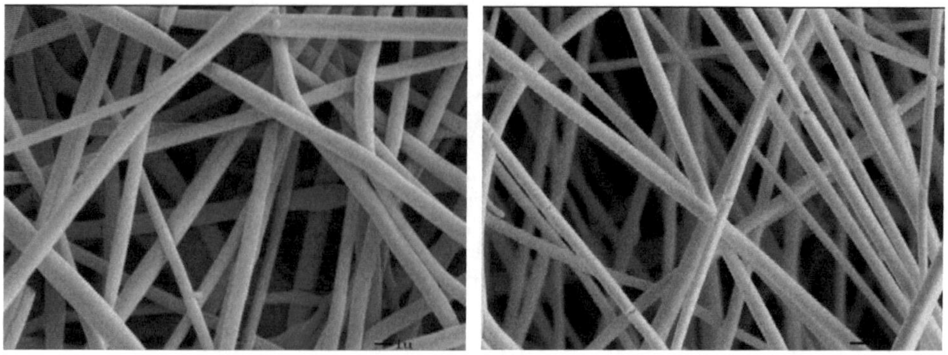

A B

Fig. 3. SEM pictures (5000x) of electrospun zein from AcOH solutions (29%) using the same spinning conditions using either commercial zein (A) or zein extracted from CGM using AcOH (B).

Chemical modification

In order to provide improved product properties, zein was allowed to react with GLY under melt conditions in a batch process as well as in an extruder. GLY has been reported to react under basic conditions with the arginine residues or the N-terminus (Marquie 2001; Miller et al. 2003). In a batch process, when a zein formulation is melt processed in the presence of water and varying levels of base with or without GLY the torque does not undergo significant increases (Figure 4). The initial increase in torque early in the run has been seen before in the presence of high amounts of water (Selling et al 2004). If GLY is not present, the resulting compression molded bars quickly dissolve in AcOH. Compression molded samples containing at least 1.5% GLY showed a measurable decrease in solubility compared to control (Table 1). The minimal differences in torque observed with the addition of GLY suggests that little crosslinking occurred during this initial mixing. In addition the dough formed immediately after melt processing only had somewhat reduced solubility. In order to determine the impact derivatization has on physical properties compression molded sample bars were produced. The heat and pressure encountered in this additional processing provides an opportunity for additional chemistry to take place which renders the sample solvent resistant. Tensile strength was not improved to the same extent as was seen when this chemistry was performed in solution. When modifying zein in the melt using GLY, tensile strength was improved by 16% when 6% GLY was used. When this chemistry was carried out in basic ethanol-water solution the tensile strength increased by over 50% (Woods and Selling 2007). This is the first demonstration of cross-linking zein in the melt state.

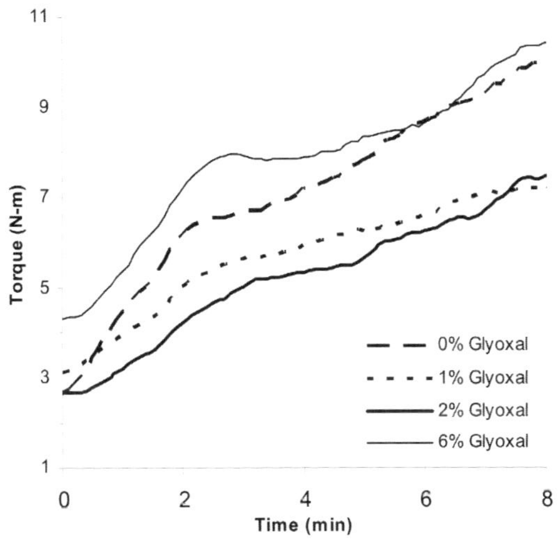

Fig 4. Effect of GLY on torque, 0.186 mmoles NaOH/gram zein, 80 °C initial temperature and 50 rpm roller rotors.

Table 1. Effect of GLY on % sample lost in AcOH after 16 hours at room temperature..

% GLY	% Loss	Swelling Ratio
0	100	N/A
1	>50	N/A
1.5	27.2	3.1
2	18	3.1
3	4	1.8
6	4.7	1.6

A series of formulations were prepared and tested to evaluate the impact of base concentration. A sample containing 6% GLY, but without NaOH, dissolves after 16 hours in AcOH. However, with a NaOH concentration as low as 0.19 µmoles/gram zein, the amount of material lost on soaking in AcOH is reduced to 25.7%. As the NaOH concentration is further increased, the percent loss plateaus at approximately 4% with the addition of at least 0.046 mmoles NaOH/gram zein. Swelling ratios generally decreased, indicative of increased cross-linking, along with the percent loss as NaOH concentration increased. Varying amounts of base also had an effect on physical properties. Only the two highest levels of NaOH (0.186 and 0.279 mmoles/gram zein) yielded samples with improved tensile strength. Other bases, such as KOH and $Ca(OH)_2$ can deliver improved solvent resistance (~ 7% sample loss) when used at 0.186 mmole/gr zein with 6% GLY, however they formulations did not provide improve physical properties. Processing conditions, such as mixing time and temperature had little impact on the solvent resistance of molded samples.

Reaction extrusion has been an effective method for derivatizing natural products in a continuous fashion (Sessa and Wing 1998; Willet and Finkenstadt 2006; Cassagnau et al 2007). Zein has been modified using anhydrides to improve metal chelation, however this adduct would still be soluble in known good solvents for zein (Sessa and Wing 1998). Zein was processed through a twin screw extruder to which was added

018.5 % of 1M NaOH base, 5.3% of a 40% GLY solution and 5.3% TEG. A control extrudate was also prepared where GLY was substituted with water. The extrudates were coarsely ground and injection molded providing sample bars that were tested for solubility. The control bars with no GLY dissolved quickly in AcOH. The injection molded containing 2% GLY had 20% loss after soaking. This is in good agreement with compression molded samples produced via the batch process.

CONCLUSION

We have developed an improved method for extracting zein from CGM utilizing AcOH in place of alcoholic solvent systems. AcOH removes zein more quickly, at lower temperatures, and higher yields when compared to alcoholic solvents. At room temperature, over 50% of the available zein is removed from CGM after 60 minutes in one pass. Yields can be improved at temperatures of 55 °C or higher. AcOH can remove protein from both GC and DDG, although the material removed from DDG has a high amount of oil in it. The purity of the zein removed from CGM is somewhat lower than commercial zein with the main impurity being fatty acids and esters. The proteinaceous component of the zein removed from CGM using AcOH as solvent is very similar to that of commercial zein. Films made from zein isolated from CGM using AcOH have lower TS and modulus when compared to films of commercial zein. Electrospun fiber can be produced from zein isolated from CGM using AcOH and is of equivalent quality as that made from commercial zein using the same spinning conditions. Improved products utilizing zein were produced using GLY as the derivatizing reagent. The effect of varying GLY, base concentration and processing conditions on compression molded zein bars produced from melt processed zein were explored. In order to improve solvent resistance, the zein required two processing steps. Melt processing was utilized to provide a homogenous blend, followed by compression molding to impart solvent resistance to samples produced using appropriate formulations. Samples containing 6% GLY with at least 0.19 μmoles NaOH/gram zein, and those including at least 1.5% GLY with 0.186 mmoles NaOH/gram zein, exhibited increased solvent resistance compared to control bars. Improved tensile strength was obtained at GLY levels of at least 3%. While the low levels of NaOH tested provided solvent resistance, the amount of NaOH had to be greater than 0.186 mmoles/gram zein to provide improved tensile strength. Other inorganic bases provided some reduction in solubility; however, tensile strength was not improved. Modification of processing conditions during the initial melt processing step had a minimal effect on the solubility of the compression molded bar. The GLY technology was carried out on an extruder where the GLY, base and plasticizer, were injected into a twin screw extruder. After injection molding, the resulting bars displayed reduced solubility in AcOH.

ACKNOWLEDGEMENTS

The Illinois Corn Marketing Board is thanked for providing financial support of this research. Ashley Maness and Gary Kuzniar are thanked for performing SDS-PAGE, microscopic, and torque rheometer experiments. Art Thompson is thanked for performing the SEM analysis. We thank Aventine Renewable Energy, Inc. and Bunge

Milling for their donations of corn gluten meal and ground corn respectively. We also thank Big River Resources for their donation of Distillers Dried Grains.

LITERATURE CITED

Biswas, A., Sessa, D. J., Lawton, J. W., Gordon, S. H., and Willett, J. L. 2005. Microwave-assisted rapid modification of zein by octenyl succinic anhydride. Cereal Chem. 82:1-3.

Biswas, A., Sessa, D. J., Gordon, S. H., Lawton, J. W., and Willett, J. L. 2005. Synthesis of zein derivatives and their mechanical properties. ACS Symposium Series 900:141-148.

Biswas, A., Shogren, R. L., Stevenson, D. G., Willett, J. L., Bhowmik, P. K. 2006. Ionic liquids as solvents for biopolymers: Acylation of starch and zein protein. Carbohydrate Polymers 66:546-550.

Cabra, V., Arreguin, R., Vazquez-duhalt, R., and Farres, A. 2006. Effect of temperature and pH on the secondary structure and processes of oligomerization of 19 kDa alpha-zein. Biochim. Biophy. Acta, Proteins Proteomics 1764:1110-1118.

Carter, R. 1970. High purity zein from low temperature solvent extraction of grain gluten. US 3,535,305.

Cassagnau, P.; Bounor, L.; Fenouillot, F. 2007. Reactive processing of thermoplastic polymers: A review of the fundamental aspects. Int. Poly. Processing 22:218-258.

Cheryan, M. 2002. Corn oil and protein extraction method. U.S. patent 6,433,146.

Cheryan, M. 2006. Method and system for extraction of zein from corn. U.S. patent 7,045,607.

Cheryan, M. 2006. Method and system for extraction of zein from corn. U.S. patent 7,148,366.

Hernandez-Munoz, P., Villalobos, R., Chiralt, A. 2004. Effect of cross-linking using aldehydes on properties of glutenin-rich films. Food Hydrocolloids 18:403-411.

Hernandez-Munoz, P., Kanavouras, A., Lagaron, J.M., Gavara, R. 2005. Development and Characterization of Films Based on Chemically Cross-Linked Gliadins. J. Ag. Food Chem. 53:8216-8223.

Kim, S., Sessa, D. J., Lawton, J. W. 2004. Characterization of zein modified with a mild cross-linking agent. Ind. Crops Prod. 20:291-300.

Lawton, J. W. 2002. Zein: a history of processing and use. Cereal Chem. 79:1-18.

Lawton, J. W. 2006. Isolation of Zein using 100% ethanol. Cereal Chem. 83:565-568.

Lee, M., Lee, S., Ma, Y., Park, S., Bae, D., Ha, S., Song, K. B. 2005. Effect of plasticizer and cross-linking agent on the physical properties of protein films. J. Food Sci. Nutr. 10:88-91.

Marquie, C. 2001. Chemical Reactions in Cottonseed Protein Cross-Linking by Formaldehyde, Glutaraldehyde, and Glyoxal for the Formation of Protein Films with Enhanced Mechanical Properties. J. Agric. Food Chem. 49:4676-4681.

Miller, A. G.; Meade, S. J.; Gerrard, J. A. 2003. New insights into protein crosslinking via the Maillard reaction: structural requirements, the effect on enzyme function, and predicted efficacy of crosslinking inhibitors as anti-ageing therapeutics. Bioorganic Medicinal Chem. 6:843-852.

Paetau, I, Chen, C. Z., Jane, J. 1994. Biodegradable plastic made from soybean products. II. Effects of cross-linking and cellulose incorporation on mechanical properties and water absorption. Environ. Poly. Deg. 2:211-217.

Parris, N., and Dickey, L. C. 2001. Extraction and solubility characteristics of zein proteins from dry-milled corn. J. Agric. Food Chem. 49:3757-3760.

Selling,G.W.; Sessa,D.J.; Palmquist,D.E. Effect of water and tri(ethylene) glycol on the rheological properties of zein. Polymer 2004, 45, 4249-4255.

Selling, G. W., Lawton, J., Bean, S., Dunlap, C., Sessa, D. J., Willett, J. L., and Byars, J. 2005. Rheological studies utilizing various lots of zein in N,N-dimethylformamide solutions. J. Agric. Food Chem. 53:9050-9055.

Selling, G. W., Hamaker, S. A. H., and Sessa, D. J. 2007. Effect of Solvent and Temperature on Secondary and Tertiary Structure of Zein by Circular Dichroism. Cereal Chem. 84:265-270.

Selling, G. W., Biswas, A., Patel, A., Walls, D. J., Dunlap, C., and Wei, Y. 2007. Impact of Solvent on Electrospinning of Zein and Analysis of Resulting Fibers. Macromol. Chem. Physics 208:1002-1010.

Sessa, D. J.; Wing, R. E. Thermochemical derivatisation of soybean and corn protein products with citric acid to enhance meatl-binding properties. In Paradigm for successful utilization of renewables, AOCS Press: Champaign, IL, 1998; pp 232-246.

Sessa, D. J.; Eller, F. J.; Palmquist, D. E.; Lawton, J. W. 2003. Improved methods for decolorizing corn zein. Ind. Crop Prod. 18:55-65.

Sessa, D. J., Mohamed, A., Byars, J. A., Hamaker, S. A. H., and Selling, G. W. 2007. Properties of Films from Corn Zein Reacted with Glutaraldehyde. J. Appl. Poly. Sci. 105:2877-2883.

Shukla, T. P. 1992. Trends in zein research and utilization. Cereal Foods World 37:225-225.

Shukla, R., Cheryan, M., and DeVor, R. E. 2000. Solvent extraction of zein from dry-milled corn. Cereal Chem. 77:724-730.

Shukla, R., and Cheryan, M. 2001. Zein: the industrial protein from corn. Ind. Crop Prod. 13:171-192.

Taylor, J.; Taylor, J. R. N.; Dutton, M. F.; de Kock, S. 2005. Glacial acetic acid - A novel food-compatible solvent for kafirin extraction. Cereal Chem. 82:485-487.

Vaz, C. M., van Doeveren, P.F.N.M., Yilmaz, G., de Graaf, L. A., Reis, R. L., Cunha, A. M. 2005. Processing and characterization of biodegradable soy plastics: Effects of crosslinking with glyoxal and thermal treatment. J. Appl. Poly. Sci. 97:604-610.

Willett, J. L.; Finkenstadt, V. L. 2003. Initiator Effects in Reactive Extrusion of Starch–Polyacrylamide Graft Copolymers. Polym Eng Sci 43:1666-1674.

Willett, J. L.; Finkenstadt, V. L. 2006. Reactive extrusion of starch-polyacrylamide graft copolymers using various starches. Journal of Polymers and the Environment. 14:125-129.

Woods, K. K.; Selling, G. W. 2007. Improved Tensile Strength of Zein Films using Glyoxal as a Crosslinking Reagent. Journal of Biobased Materials and Bioenergy. 2:281-287.

Wu, Q., Hiroshi, S., and Seiichiro, I. 2003. Studies on the toughness and water resistance of zein-based polymers by modification. Polymer 44:3901-3908.

Wu, Q., Yoshino, T., Sakabe, H., Zhang, H., and Isobe, S. 2003. Chemical modification of zein by bifunctional polycaprolactone (PCL). Polymer 44:3909-3919.

Yang, Y., Wang, L., Li, S. 1996. Formaldehyde-free zein fiber. Preparation and investigation. J. Appl. Polym. Sci. 59:433-441.

280b Properties Of Reactively-Mixed Molded Dairy-Based Biocomposites

C.I. Onwulata, Dairy Processing and Products Research Unit, USDA-ARS Eastern Regional Research Center, 600 E. Mermaid Lane, Wyndmoor, PA 19038 and P.M. Tomasula, USDA, ARS, Eastern Regional Research Center, USDA-ARS Eastern Regional Research Center, 600 E. Mermaid Lane, Wyndmoor, PA 19038.

The increasing demand for environmentally friendly/biodegradable plastics has driven research to explore the use of agriculturally-derived materials in biocomposites. Dairy proteins have functional properties similar to many plastics and may be suitable replacements for plastics in biocomposites; however, they are not as strong. We blended whey protein isolates and casein (40%), corn starch (30%), glycerol (10%), cellulose (5%), and water (15%) with Acetic Acid (AA) in varying ratios to form the dairy biocomposite. The blend was mixed with roller blades in a torque rheometer for 40 min at initial temperature of 70 °C. After mixing, the biocomposite paste was molded into ASTM stripes at 121 °C for 30 min using a Carver press. The results showed that as the ratio of AA increased, the storage modulus of the biocomposite increased to an optimum (862.1 MPa) at 3.75% AA compared to the control (676.0 MPa) with 0% AA. This result shows that reactively mixing dairy biocomposites with AA improves mechanical strength.

Studies on Mechanical Properties and Enzymatic Degradation of Partially Biodegradable Laminated LDPE - Starch Blend Film

Asim K. Jana, *Department of Chemical and Bioengineering, National Institute of Technology, Jalandhar, India,* ***Sangeeta Garg***, *Department of Chemical and Bioengineering, National Institute of Technology, Jalandhar, India and* ***Sarabjit Singh***, *Department of Applied Physics, National Institute of Technology, Jalandhar, India*

**Corresponding author mailing address*
Dr. A. K. Jana; Tel: 91-181-2690301 Ext 393; E mail: janaak@nitj.ac.in

Introduction

Plastic is one of the largest groups of polymeric materials of petrochemical-based organic origin formed into useful shapes by application of heat and pressure. The most important of them are Polyethylene (PE), Polypropylene (PP), Linear Low Density Polyethylene (LDPE) and Polyvinyl Chloride (PVC). These are not environment friendly as they take years to decompose [1] and also the petroleum resources are depleting day by day. The development of modified plastics and substitutes that could be degraded by micro organisms in soil and water has been utmost necessity with the increasing volume of industrial and the municipal waste.

Starch is a cheap natural biodegradable polymer of carbohydrate available abundantly from plant source. Plastic starch blends can be used as an alternative to the plastics [2-7]. The biodegradation of starch from the blend film in the environment causes deterioration of the matrix, leading to the formation of a polymer mesh/ small pieces, which accelerate the disintegration and degradation of the plastic materials.

The blending of the hydrophilic starch and hydrophobic LDPE are not compatible in nature and have poor surface adhesion leading to phase separation during the processing of the film. The mechanical properties of the films deteriorate rapidly with the increase in starch concentration [8-12]. To improve the quality of the blend film various modifications of the starches have been tried but with limited success [5, 7, 8, 12]. In the present work, with the objectives to enhance the mechanical properties at higher starch content i.e. improved biodegradation properties, laminated films of the LDPE-starch blend film laminated on both sides of the LDPE films have been prepared. The mechanical properties and enzymatic biodegradations of the films have been studied.

Experimental

Low Density Polyethylene (LDPE) used in this study was made by Indian Petrochemical Corporation Limited, Baroda and Starch (corn starch) was made by Sukhjit Starch and Chemicals Limited, Phagwara (India). Glycerol used was reagent grade chemical. Modified starch was prepared by treatment of glycerol with the starch in a Tumbler mixer. A batch is charged into the mixture to 50 % full and the ends are closed. The partly filled mixture was rotated about the horizontal axis with 30 rpm speed. The time of mixing for each batch of feed was 15 mins.

Blend films were prepared by incorporating starch, glycerol modified starch in LDPE. The blend mixture was prepared by taking the different components in appropriate ratios and films were made by extrusion and blow film plant (Konark Instruments Ltd, India). Composition of the various films prepared is summarized in Table 1.

Table 1: Composition of LDPE - starch blend films.

Film	Composition (% w/w)		
	LDPE	Starch	Glycerol
LDPE	100	-	-
LDPE–10NS	90	10	-
LDPE-10 MS	80	10	10
LDPE-15 MS	75	15	10

LDPE - Low -density polyethylene; NS - Native starch; MS - Modified starch i.e. starch modified with glycerol.

For the preparation of the laminated film, two LDPE – starch blend films were laminated on both sides of LDPE film. It was prepared by passing the three layers of films through hot press rollers rotating in opposite directions to each other. The temperatures of rollers were 150 ^{o}C. The compositions of laminate films are shown in Table 2.

Table 2: Compositions of laminate films

Laminates	Composition (% w/w)		
	LDPE	Starch	Glycerol
Laminate – I	86.66	6.67	6.67
Laminate – II	83.33	10	6.67

Tensile strength and percent elongation of the films were determined by Instron Universal Testing Machine (Lloyd instruments, UK) at load cell 100 N, load range 5 and cross head speed 100 mm/min. Tensile strength and % elongation of the films were determined per unit thickness of the film. Measurements were done in 5 replicates.

In the enzyme biodegradation of the films, thermostable α-amylase was produced using the *Bacillus licheniformis* and the extracellurly produced enzyme in the fermentation broth was directly applied for the degradation studies. Films have been characterised before and after the degradations.

Results and Discussion

Hydrophilic starch is incompatible with the hydrophobic LDPE, which results in poor quality blend. Blending of starch with LDPE drastically reduces the mechanical properties of LDPE - Starch blend films and it is also difficult to increase `the concentration of starch in the blend. The reduction of hydrophilicity of the starch by starch modification could solve the above problem. Concentration of starch also could be increased by producing laminates with two LDPE - Starch blend films on both sides of LDPE film.

The results have been presented in the following major points

I. Processing of LDPE - starch blend films and laminates
II. Mechanical properties of LDPE - starch blend films and laminates.

Processing of Films and Laminates

It was observed during the processing of films that the thickness of LDPE - starch blend films were larger than that of LDPE film. The smoothness of starch blend films were also less than that of LDPE film. As the percentage of starch was being increased in the

LDPE films, processing was becoming more and more difficult as the films were getting punctured. It was also observed that the thickness of the films could be controlled by the air blown through the die orifice and by the speed of the rollers pulling the film. As the starch content was increased in the films, film was becoming more and more non-uniform due to low compatibility of starch in LDPE. So, glycerol was added to increase the uniformity and to decrease in the roughness in the film. After adding the measured quantity of glycerol, another difficulty in the processing of the films was observed. The material was not flowing freely through the extruder. So after the addition of glycerol in the LDPE - starch mixture, the whole of the mixture was kept in the oven for 30 minutes at a temperature of 50 °C and then processed in the usual manner.

It was observed during the processing of laminate films that laminates could also be prepared by compression molding technique, but continuous laminates could not be prepared by compression molding and also it was time consuming. So, the laminates were prepared by hot pressing through the heated rollers. While passing through the heated steel rollers the films to be joined were covered with a thick paperboard to avoid laminates from sticking to the surface of the rollers.

Mechanical Properties

Mechanical properties of tensile strength and percent elongation were studied.

Tensile strength

Tensile strength of various LDPE - starch blend films and laminates are shown in Figure 1. It was observed that the tensile strength of all the films in lengthwise were higher

Fig.1. Tensile strength of LDPE - starch blend films and laminates

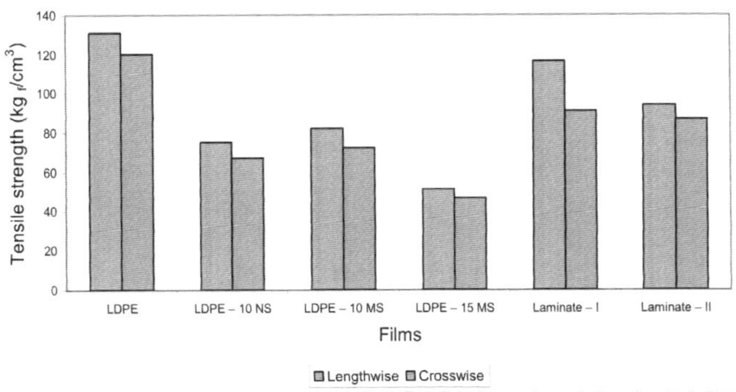

than tensile strength in crosswise. The trends in variation of tensile strength in lengthwise with respect to strength in crosswise were similar. From the results, it was observed that tensile strength of LDPE film was the maximum amongst all blend film samples and it was 131.2 Kg_f / cm^3. Tensile strength decreased to 75.31 Kg_f / cm^3 on the addition of 10 % native starch in LDPE – 10 NS blend film. Tensile strength further decreased to 51.35 Kg_f / cm^3 in blend film with 15 % modified starch (i.e. LDPE - 15 MS blend film). It was least amongst all the blend films. Tensile strength of the Laminate - I (two layers of LDPE - 10 MS on both sides of LDPE) was measured to be 116.6 Kg_f / cm^3 while tensile strength of Laminate - II (two layers of LDPE - 15 MS on both sides of LDPE) was decreased to 93.98 Kg_f / cm^3.

It was clear that with the addition of starch in the blend films, tensile strengths decreased. LDPE is hydrophobic and starch is hydrophilic. So starch showed a poor compatibility with LDPE. It may be also due to the incorporation of starch decreased the molecular orientation of LDPE and increased the stiffness of the films, which ultimately reduced the mechanical integrity of the blend films.

In case of laminates, tensile strengths of laminate films were higher than tensile strengths of all blend films prepared in this study but were still less than tensile strength of LDPE film. The tensile strength of Laminate - I was higher than tensile strength of Laminate - II because of lower starch content in Laminate - I (6.67%) than in Laminate - II (10%).

Variation in tensile strength with starch content in different blend films and laminates is shown in Figure 2. Curves were drawn for LDPE - native starch blend films, LDPE - modified starch blend films and laminates.

Fig. 2. Variation of tensile strength with starch content in different blend flms and laminates

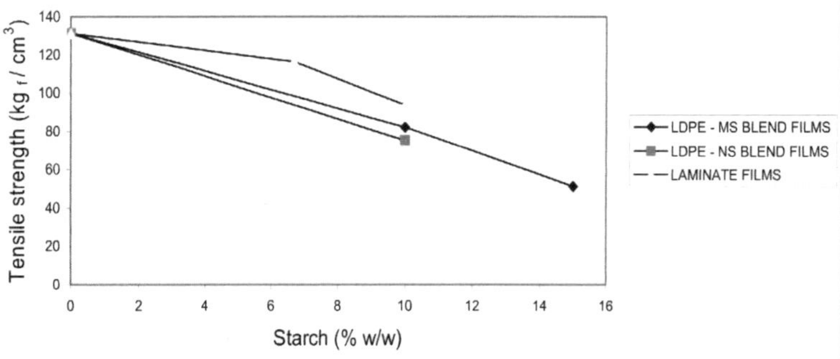

These curves clearly showed that tensile strength decreased with increase in starch content of blend films and laminates. However the extent of decrease depended on the concentration of starch in the blends and laminates. Tensile strengths of laminate films were less than tensile strength of LDPE film but higher than the tensile strengths of all other LDPE - starch blend films at the corresponding starch concentrations.

It was also observed that tensile strength of blend films with modified starch was higher than tensile strength of blend films with native starch. Effect of glycerol modification of starch on tensile strength of films was compared by considering LDPE - 10 NS and LDPE -10 MS blend films. Tensile strength of LDPE - 10 MS film with 10 % modified starch was higher than tensile strength of LDPE - 10 NS blend film with 10 % native starch. These results could be explained on the basis of changes in starch properties by modification. Native starch granules exhibit hydrophilic properties and strong inter-molecular association via hydrogen bonding due to hydroxyl groups on the granule surface. The strong hydrogen bonding association and crystallization leads to poor thermal processing and poor compatibility with LDPE blend. The changes in starch properties by modification affected the compatibility of the blend in the mixture. The compatibility of the blend of starch and LDPE increased. So, LDPE films containing modified starch showed higher mechanical strength than the films containing native starch.

Percent elongation

Percent elongation is the ability of the films and laminates to stretch. Percent elongation of various LDPE - starch blend films and laminates are showed in Figure 3. It was observed that percent elongation of LDPE was the maximum among all samples and it was 5023.45 %. Percent elongation decreased to 1012.12 % on the addition of 10 %

Fig. 3. Percent elongation of LDPE - starch blend films and laminates

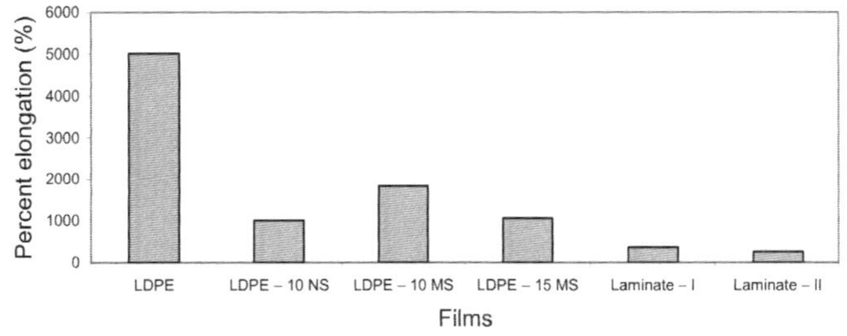

starch i.e. in case of LDPE - 10 NS blend film. In LDPE - 10 MS blend film, percent elongation was measured to be 1837.43 %, which was greater than percent elongation of LDPE - 10 NS blend film but was less than percent elongation of LDPE film. When starch was taken to be 15 % in LDPE - 15 MS blend film, percent elongation was further decreased to 1066.56 %, which was very close to percent elongation of LDPE - 10 NS film.

Percent elongation of the Laminate - I (with two layers of LDPE - 10 MS blend films laminated on both sides of LDPE film) was measured to be 368.10 % while percent elongation of Laminate - II (with two layers of LDPE - 15 MS blend films and middle layer of LDPE film) was measured to be 256.00 %. It was observed that, percent elongation of laminate films were much less than those of all other blend films.

The results showed that percent elongation of blend films decreased with the addition of starch due to lower compatibility of starch with LDPE.
In case of laminates, percent elongation of laminate films were lower than percent elongation of LDPE film and all blend films prepared in this study. This was due to the stiffness of laminate films. Starch content was much higher in laminate films than all other blend films, which might have decreased molecular orientation and increase in their stiffness. Also percent elongation of Laminate - I was higher than percent elongation of Laminate - II because of lower starch content in Laminate - I (6.67%) than in Laminate -II (10%).

Variation in percent elongation with starch content in different blend films and laminates is shown in Figure 4.

Fig. 4. Variation of percent elongation with starch content in different blend films and laminates

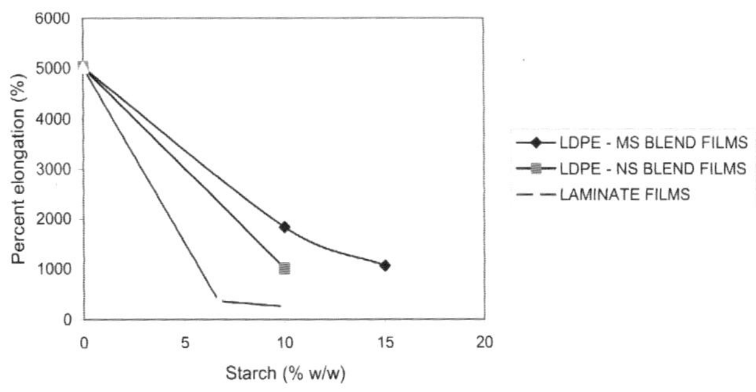

Curves were drawn for LDPE - native starch blend films, LDPE - modified starch blend films and laminates separately. It clearly showed that percent elongation decreased with increase in starch content of blend films and laminates. However, the extents of decrease in case of laminates were much larger than the blend films at the corresponding starch concentration. It was also observed that percent elongation of blend films with modified starch was higher than percent elongation of blend films with native starch.

Effect of glycerol modification of starch on percent elongations of the films were studied by considering LDPE - 10 NS & LDPE -10 MS blend films. Percent elongation of LDPE - 10 MS film with 10 % modified starch was higher than percent elongation of LDPE - 10 NS blend film with 10 % native starch. This was due to the fact that with the addition of plasticizer glycerol, enhanced the compatibility of LDPE and starch blend which increase in the flexibility of blend film containing modified starch.

In the enzymatic degradations of the films, results showed that starches in the blend films were degraded by the enzyme but rates varied with accessibility of starch to the enzyme. Characteristics of the films before and after degradation were evaluated by mechanical properties. Reduction in tensile strength and % elongation depending upon starch content of films were observed after degradations. Reduction increased with the increase in starch content of the films. The tensile strength and % elongation of blend film containing 15 % starch reduced by 9.5 % and 14 % respectively in 72 hours of time. The activity of the enzyme decreased considerably after 72 hours under experimental conditions.

Conclusions

Increase in starch content deteriorates the mechanical properties of LDPE – starch blend films. Starch is hydrophilic while LDPE is hydrophobic in nature. When starch is blended with LDPE, mechanical properties of the films deteriorate due to low compatibility of starch with LDPE. When starch was modified by adding glycerol (plasticizer), glycerol plasticizes effectively to reduce internal hydrogen bonding in starch. It enhanced the compatibility of starch with LDPE and increased the flexibility of the blend film.

Two starch blend films were laminated on both sides of LDPE film. The starch contents of the laminated films were higher than the blend films. The mechanical properties were measured. With the addition of starch, there was a decrease in tensile strength and percent elongation in laminates. Tensile strengths of laminated films were found to be much higher than any individual blend film. But there was a decrease in its percent elongation. The extent of decrease in tensile strength with the increase in starch concentration was smaller in the laminated film than the blend films.

Biodegradation of the films were studied by the enzyme α-amylase producod from *Bacillus licheniformis*. Reduction in tensile strength and % elongation depending upon starch content of films were observed after degradations.

References

1. Potts, J.E., Clendinning, R. A., Ackart, W.B., & Niegisch, W. D. (1973). Polymer and ecological problems. New York: Plenum Press, p.61.
2. Chandra, R., & Rustgi, R. (1998). Biodegradable polymers. Progress in Polymer Science, 23, 1273 – 1335.

3. Raj, B., Raj, E., & Kumar, K., (2002). Moisture –Sorption Characteristics of Starch/ Low – Density Polyethylene Films. Applied Polymer Science, 84, 1193-1202.
4. Otey, F.H., & Westhoff, R.P. (1984). Starch-based films. Preliminary diffusion evaluation. Industrial Engineering Chemistry Production Research and Development, 23, 284 - 287.
5. Soundararajan, S., Palanivalu, K., & Sharma, S.K. (2002). Studies on starch filled PE (film) and PP (sheet). Plastic processing and uses, 47, 80-83.
6. Kim, M. (2003). Evaluation of degradability of hydroxypropylated potato starch / polyethylene blends films. Carbohydrate Polymers, 54, 173-181.
7. Kim, D., Ki Na, S., & Park, J., (2002). Preparation and Characterization of modified starch based Plastic film reinforced with short pulp fiber II. Mechanical Properties. Journal of applied polymer Science, 88, 2108 – 2117
8. Kim, M., & Lee, S. (2002). Characteristics of cross-linked potato starch and starch filled linear low-density polyethylene films. Carbohydrate polymers, 50, 331-337
9. Kuniak,L., & Marchessault, R.H. (1972). Study of the cross linking, reaction between epichlorohydrin and starch. Starch/Strake, 24, 110-116.
10. Franck, D., Gregorio, C., Sabrina, B., Claudine, F., & Giangiacomo, T. (2004). Preparation, characterization and sorption properties of cross linked starch – based exchangers. Carbohydrate Polymers, 60, 67-75.
11. Chung, H., Soo Woo, K., & Lim, S., (2004). Glass transition and enthalpy relaxation of cross-linked corn starches. Carbohydrate Polymers, 55, 9 - 15.
12. Wurzburg, O. B. (1987). Cross – linked starches. In O. B. Wurzburg (Ed.) Modified starches: Properties and uses. Boca Raton, FL: CRC Press. 46 - 51.

280d Biodegradable Green Hybrid Composites From Bacterial Bioplastic

Sanjeev Singh and A.K. Mohanty. School of Packaging, Michigan State University, 130 Packaging Building, MSU, East Lansing, MI 48824

Hybrid green composites from a biopolymer, wood fiber and talc were successfully fabricated through extrusion followed by injection molding processing. The fabricated green composites were characterized for their mechanical properties. The biodegradable plastic used in this study is a bacterial biopolyester, polyhydroxybutyrate-co-valerate (PHBV). The wood fiber and talc hybrid reinforcements of PHBV showed significant improvement in properties of the virgin bioplastic. The hybrid composite was targeted to design a balance in the properties with their cost effectiveness and environment friendliness. One of the targets of this investigation is to establish the structure-property-processing co-relationship of the hybrid composites. The potential of these developing green composites in automotive parts and rigid packaging will be highlighted and discussed.

280e Biocomposite Sheet Molding Compound Processing for Production of Auto Parts to Housing Structures

Manju Misra, Chemical Engineering and Materials Science, Composite Materials and Structures Center, Michigan State University, 2100 Engineering Building, East Lansing, MI 48824 and Lawrence T. Drzal, Composite Materials and Structures Center and Chemical Engineering and Materials Science, Michigan State University, 2100 Engineering Building, East Lansing, MI 48824-1226.

Natural/Bio-fiber composites (biocomposites) are emerging as a viable alternative to glass fiber reinforced composites especially in automotive and building product applications. The combination of cellulosic biofibers such as kenaf, industrial hemp, flax, jute, henequen, pineapple leaf fiber(PALF) and sisal with polymer matrices from both non-renewable and renewable resources to produce composite materials that are competitive with synthetic composites requires special attention i.e. biofiber-matrix interface and novel processing. This presentation will highlight on biocomposite sheet molding compounding (BSMC) processing technology to fabricate structural and semi-structural composites for automotive parts and housing structures.

PHYSICO-MECHANICAL AND MORPHOLOGICAL STUDY ON HEMP-PBS BIOCOMPOSITES

B. N. Dash, M. Nakamura, S. Sahoo, M. Kotaki, A. Nakai and H. Hamada
Department of Advanced Fibro-Science, Kyoto Institute of Technology
Matsugasaki, Sakyo-Ku, Kyoto-606-8585, Japan
Phone: +81(0)75-724-7844
Fax: +81(0)75-724-7800
Baikuntha.dash@gmail.com, i2ufc105@kit.ac.jp,
saswata.sahoo@gmail.com, m-kotaki@kit.ac.jp,
Nakai@kit.ac.jp, hhamada@kit.ac.jp

Abstract

Most commercially available plastics today are non-biodegradable and cause serious concern for the environment, associated with their disposal. Hence the use of renewable resources like biopolymers and biofibers for the fabrication of biodegradable plastics is the need of the hour that can partially solve the problem of plastic pollution. In the present research biofiber hemp (untreated and surface modified) has been effectively reinforced with biopolymer PBS (Polybutylene succinate) to fabricate biocomposites using injection molding method. Surface modification of hemp fiber was done with aqueous shellac resin to study the characteristics of the treated hemp-PBS composite. Fiber loading up to 20 wt% with PBS could be achieved. Mechanical properties (Tensile and Izod impact), moisture absorption analysis and SEM (Scanning electron microscopy) analysis were done to evaluate the properties of the biocomposite. Surface treatment with shellac resin resulted in an increase in tensile modulus and decrease in tensile strength as compared to untreated hemp composites. The impact strength of composite specimens (with and without notch) decreased with increase in fiber loading in comparison to neat PBS although the treated composite specimens had better impact strength than the untreated specimens.

Key Words:- PBS, Hemp, Biodegradable, Surface modification, Biocomposite

Introduction

Most commercially available plastics today are non-biodegradable and cause serious concern for the environment, associated with their disposal. Hence the use of renewable resources like biopolymers and biofibers for the fabrication of biodegradable plastics is the need of the hour that can partially solve the problem of plastic pollution. Biocomposites, prepared by reinforcing bio-fibers with biopolymers at low cost with acceptable mechanical properties are being developed which may be a viable substitute of the non-biodegradable plastic composites[1].Natural fibers are renewable, abundant, eco-friendly, inexpensive and possess higher specific strength properties than synthetic fibers while biopolymers can maintain the sustainable development of an economically and ecologically attractive technology. However the interfacial adhesion between the biofibers and biopolymers is not adequate because the biofibers are highly hydrophilic in nature, aggregate under processing temperature and are not homogeneous like synthetic fibers. They contain a lot of components like cellulose, hemicellulose, lignin, wax etc[2,3].Therefore several workers have reported different types of surface modifications of the biofibers for imparting strength, homogeneity, degumming, moisture repellence, flame retardant properties etc. to enhance the interfacial adhesion between the components of the biocomposite[1,3]. In the present work, biocomposites from PBS matrix reinforced with untreated and shellac treated hemp fiber have been fabricated by injection molding method and characterized.

Experimental

Materials

PBS (Polybutylene succinate)

PBS is one of the ductile thermoplastic aliphatic polyesters with a range of desirable properties including biodegradability; melt processability along with thermal and chemical resistance[4].It is a semi crystalline polymer, prepared through the condensation reactions of glycols such as butane-1, 4-diol and aliphatic dicarboxylic acid such as succinic acid as the main starting materials. It degrades in the natural environment by bacteria and fungi[4]. The structure of PBS is as follows[5].

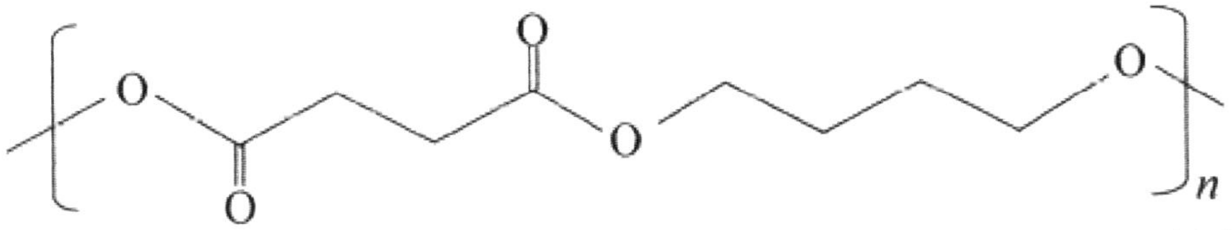

PBS has a large number of methylene groups with a planar zigzag structure which resembles PE(polyethylene).It has got limited ecological applications like preparation of mulch films, delivery systems for fertilizers and pesticides, disposable packaging, refuge bags etc.[6].
One of the possibilities to diversify the use of PBS is to reinforce it with fibers to enhance its mechanical and thermal stability for specific end uses. Recently attempts are being made to fabricate biocomposites by reinforcing agro fibers with PBS matrix[5, 7-9].In this work, PBS pellets (Grade-AZ91T), supplied by Mitsubishi Chemicals Co. Japan, have been incorporated with hemp for the fabrication of biocomposites. The pellets have a melt flow index of 4.5 g/10 min (190^{0}C/2.16 kg), a density of 1.26 g.cm^{-3} and melting temperature of 110^{0}C.

Hemp fiber

Hemp is a distinct variety of cellulosic bast fiber belonging to the plant species cannabis sativa L. Hemp fiber contains mainly cellulose (55-77 %), hemicelluloses (8-19 %), Lignin (2-5%) together with some wax and minerals[10].Growing and harvesting conditions (nature of soil, weather conditions, maturity, fertilization, portion of the plant used as fiber etc.) influence its yield, fiber strength, chemistry and surface morphology[11].The major drawback of hemp is its hydrophillicity (13 %)[7].This significantly reduces the fiber-matrix adhesion. So in this work, surface modification of chopped industrial hemp fibers (supplied by Japan Ecology Production Co.) has been done with aq. shellac solution in an economical way.

Shellac

Shellac is a brittle, flaky, nonpoisonous natural organic biodegradable resin that comes from an insect, Laccifera lacca. Shellac is chemically similar to synthetic polymers and hence called natural plastic. It can be molded by the application of heat and pressure for which it is classified as a thermoplastic. It is characterized by its high adhesive strength, excellent hardness, great electrical insulation property, high plasticity and good dye fixing behaviour [12]Each molecule contains four or five hydroxyl groups along with one free carboxyl group, one aldehyde group, two or three ester radicals and one unsaturated bond[2,13].The shortcomings of shellac are its brittleness, low softening point and bad water resistance etc.[12].Shellac is soluble in various organic solvents especially in alcohol as well as in alkaline solutions such as ammonia, sodium hydroxide etc. When dissolved in denatured alcohol, it yields a coating of superior durability and hardness. In the present investigation, 2 wt.% aq. solution of shellac (grade-LB-51) has been used, supplied by Gifu Shellac Ltd. Japan. Shellac is insoluble in water. So it was dissolved in ammonia and then diluted with water to prepare 2 wt.% aq.solution.

Surface treatment of hemp fiber

The hemp fibers were soaked in 2 wt.% aq. solution of shellac[2] for 10 minutes and then transferred to a plastic sieve to drain out the excess solution. The treated fibers were air dried for 12 hours followed by oven drying for 12 hours at 80^0 C. Finally the fibers were oven dried for 1 hour at 100^0C to remove moisture if any before compounding. In the present discussion shellac treated fibers have been represented as treated.

Fabrication of biocomposite

It was not possible to compound the randomly oriented hemp fibers with PBS pellets directly because fibers aggregate under processing temperature. So PBS sheets (Fig. 1 a) were prepared by compression molding using a 50 ton hydraulic press at 135^0C and 40 kgf. pressure. Fibers were put inside two PBS sheets in a sandwitched pattern (Fig.1 b) and again pressed in the same press under identical conditions for wetting of the fibers with PBS (prepreg). These sheets were then crushed. In case of 20 wt % loading, wetting of the fibers by the matrix was not satisfactory after the first press because of higher volume of fiber and fiber agglomeration was observed. So the crushed materials were repressed in the compression mold and crushed. These crushed materials were then compounded in a twin screw extruder (ISUZU-JSW TEX 30 HSS) at 135^0C, to get the composite pellets. Fiber content (both untreated and shellac treated) was varied to achieve 10, 15 and 20 wt. % loading of fiber. The neat polymer was used as a control. These composite pellets were then molded into test specimens by an injection molding machine (PO YUEN, TM 30H,Toyo Machinery & Metal Co,ltd. Germany). The barrel temp was set at 135^0C.

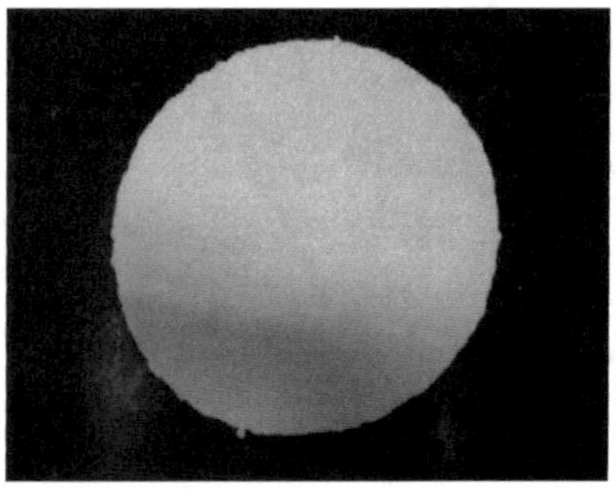

Figure 1a- PBS sheet.

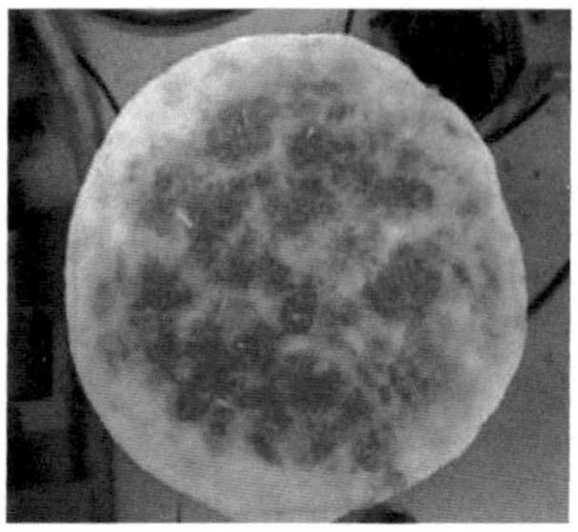

Figure 1b- Sandwiched sheet.

Characterization of composites

Tensile test and flexural strengths (three point bending) were carried out with a universal testing machine (Instron-4466) at room temperature, taking five specimens (span length-115mm) of each category followed by their average values. The gauge length was 115 mm and the crosshead speed was10mm/min. Izod impact strength (with and without notch) of the composite specimens was measured in a Digital Impact Tester (TOYOSEIKI) at room temperature. The impact strength of only notched PBS specimens was determined as PBS is ductile. The hammer energy was 5.5 Joules. Moisture absorption characteristics of untreated and treated hemp fiber were also determined at room temperature. Morphology of the untreated as well as shellac treated fibers along with tensile fractured surfaces of PH-20 wt %

and PHS-20 wt% specimens were studied using a scanning electron microscope (SEM) (JSM-5200, JEOL Ltd. Japan).

Results and Discussion

Figure 2a shows the tensile modulus of the different specimens (neat PBS, PH-10 wt.%,PH-15 wt.%, PH-20wt.%,PHS-10wt.%, PHS-15wt.% and PHS-20 wt.% where P stands for PBS, H represents untreated hemp & HS represents shellac treated hemp fiber). Figure 2b shows the tensile strength of different specimens. The dotted lines in the figures represent neat PBS. From Fig. 2a, it is clear that tensile modulus increases with increase in fiber loading for both types of composite specimens. Tensile modulus of 10 wt.% shellac treated specimens(1.30 GPa) were found to be better than that of 10 wt.% untreated specimens(1.09 GPa), while it was almost same for other two types of fiber loading i.e., 15 wt.%(1.67 GPa) and 20 wt.%. All types of composite specimens showed significant increase in tensile modulus than neat PBS (0.69GPa). PH-20 and PHS-20 wt.% specimens gave the best result (2.09 GPa) i.e., an increase of about 200 % in comparison to the neat polymer. Increase in fiber loading resulted in an increase in the stiffness of the materials while surface treatment of the fibers with shellac might have resulted in more stiffness because of its adhesive properties. This can be observed from the SEM photographs (Fig. 3a & b) where as a thin film of shellac resin is seen on the surface of the treated fiber making the individual filaments bundled and giving it a smooth and stiff surface. In case of untreated fiber, the surface is rough and individual filaments are clearly seen. The difference in the modulus values of both types of specimens was not more

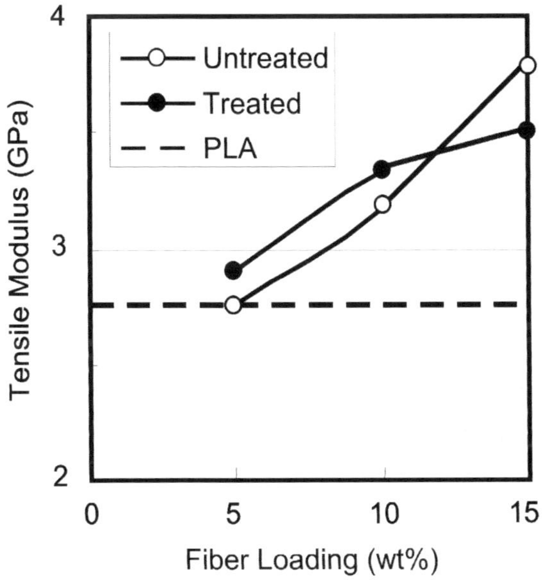

Figure 2a Tensile strength of Hemp/PLA specimens.

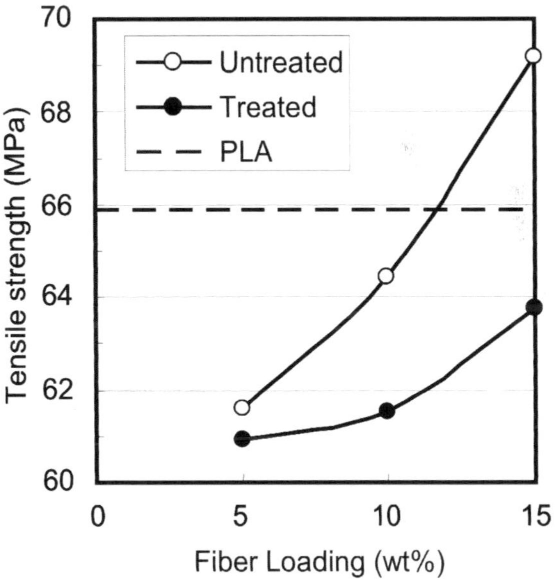

Figure 2b Tensile modulus of Hemp/PLA specimens.

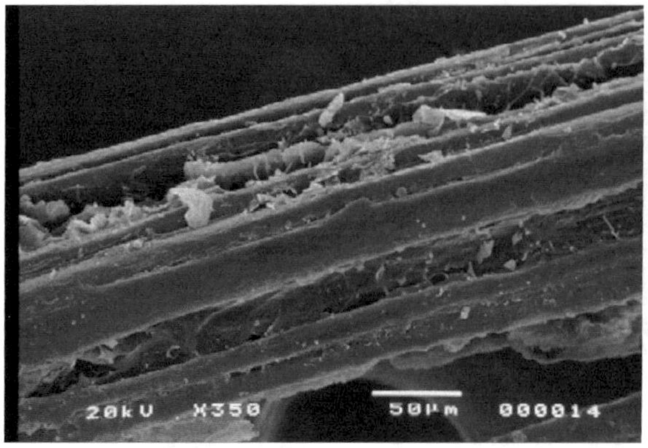

Figure 3a SEM photographs of control hemp fiber.

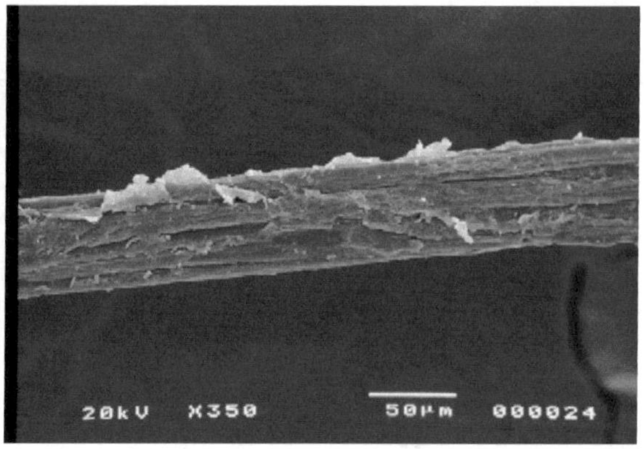

Figure 3b SEM photographs of Shellac treated hemp fiber.

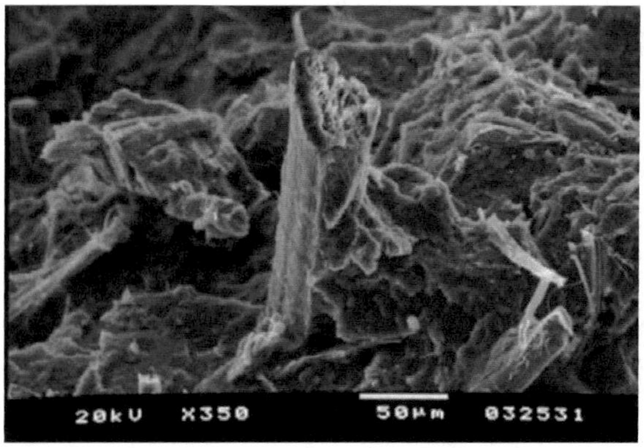

Figure 4a SEM photographs of control hemp/PLA fractured

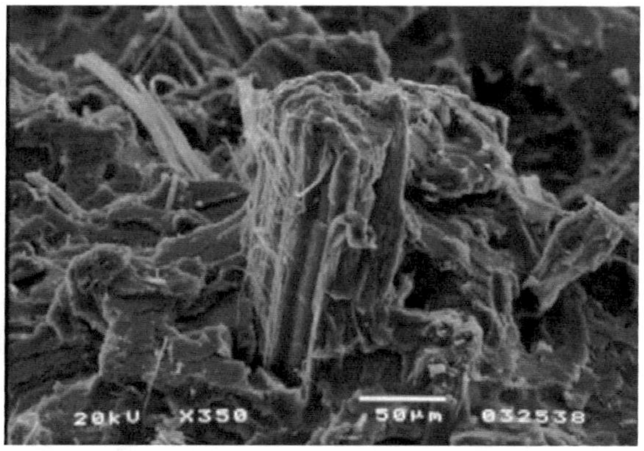

Figure 4b SEM photographs of Shellac treated hemp/PLA fractured

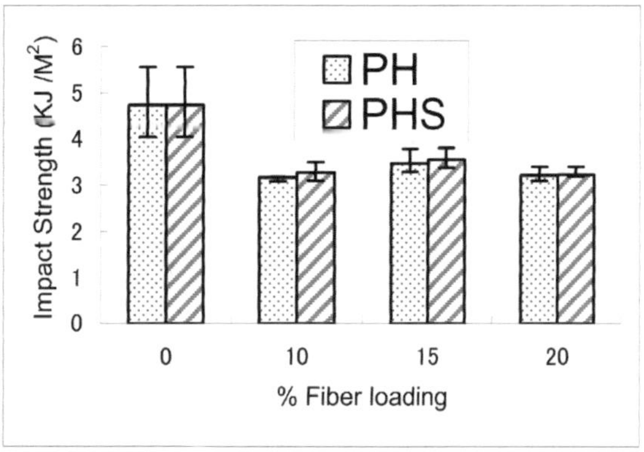

Figure 5a Impact strength of composites (notched)

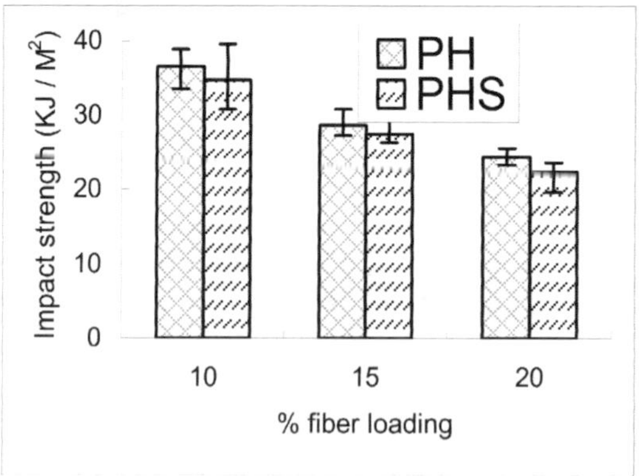

Figure 5b Impact strength of composites (unnotched)

because of very low concentration of shellac treatment (2 wt.%). Tensile strength of all types of composite specimens (Fig.2b) was better than neat PBS (36.2 MPa). The untreated specimens had in general better strength than treated specimens. PH-20 wt% showed the highest tensile strength (39.4 MPa), followed by PHS-20 wt% i.e., the stress could be easily transferred from the matrix to the untreated hemp fibers than the treated stiff fibers in the composite specimens, accounting for better interfacial adhesion. SEM photographs of the fractured surfaces of PH-20 wt.% (Fig. 4-a) shows that there is better adhesion between the fiber and the matrix, where as SEM photographs of the fractured surfaces of PHS-20 wt.% (Fig. 4- b) shows a gap between fiber and matrix.

Impact strength curve (Fig.5-a) of un-notched composite specimens shows that with increase in fiber loading, impact strength decreases as the sudden impact energy is mostly carried by the matrix resin(major component) than the fiber. The same trend was observed in case of notched specimens (Fig.-5-b) having highest value for neat PBS (4.736 J/m^2).Surface treated specimens showed relatively better impact strength than untreated specimens.

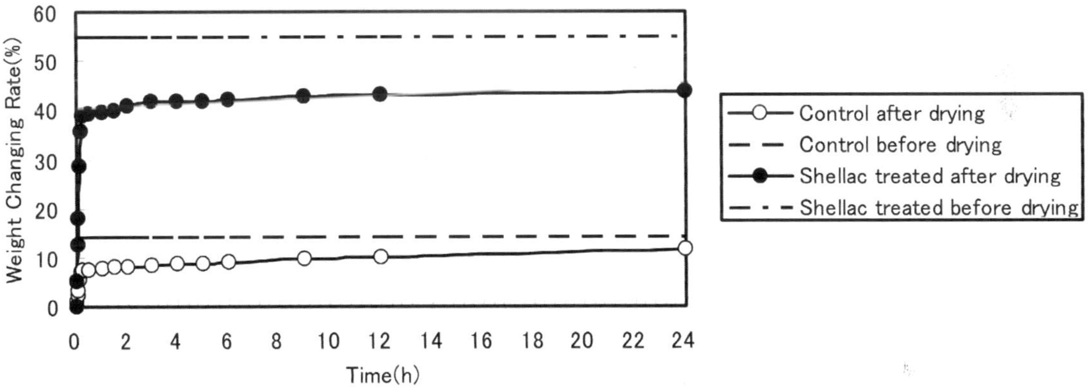

Figure 6 Moisture absorption curves of control and shellac treated hemp fibers.

Figure 6 shows the moisture absorption curves of both types of hemp fibers. The dotted lines represent the moisture contents of untreated and treated hemp at room temperature before drying. Moisture absorption data from Fig. 6 shows that shellac treated hemp fibers absorbs appreciable amount of moisture (almost three times) than that of untreated hemp within 24 hours at room temperature. High moisture absorption of the shellac treated fibers might have affected its compatibility with PBS.

Conclusion

Hemp fiber could be effectively reinforced with PBS matrix to fabricate biocomposites with better tensile and flexural properties. Fiber loading upto 20 wt.% could be achieved with the matrix by injection molding method. Chemical modification of hemp fiber with aq. shellac resulted in an appreciable increase in tensile modulus of the biocomposites. Surface modification of hemp with alcoholic solution of shellac may be more effective for achieving still better strength and moduli.

References

1. A. K. Mohanty, M. Misra and L. T. Drazl., Ch.1, Natural Fibers, Biopolymers and biocomposites., Taylor and Francis Group, New York, (2005).

2. M. Nakamura, K. Kitagawa, S. Sahoo, U. S. Ishiaku, A. Nakai, M. Kotaki and H. Hamada., Proceedings of the 4th International workshop on Green Composites., Tokyo, Japan, (2006).
3. A. Bismarck, S. Mishra and Thomas Lampke.,Ch.2., Natural Fibers, Biopolymers and biocomposites ,Taylor and Francis Group, New York, (2005).
4. T. Fujimaki, Polym. Degrad. Stab., **59,** 209 (1998).
5. T. Ohkita and seung-Hwan Lee., J. Appl. Polym.Sci. **97,** 3 (2005).
6. Y. Ikada and H. Tsuji., Macromol. Rapid Commm., **21**, 117-132 (2000).
7. Hee-Soo Kim, Han-Seung Yang and Hyun-Joong Kim., J.Appl.Polym.Sci. **97**,4(2005).
8. M. shibata, S. Oyamada, Shin-ichi Kobayashi and D. yaginuma., J. Appl. Polym. Sci. **92**, 3857-3863(2004).
9. S. M. Lee, C. Donghwan, W.H. Park, S. G. Lee, S.O.Han and L. T. Drazl., Compos. Sci.Technol., **65**, 647-657 (2005).
10. A. B. Thomsen, V. Bohn, K. V. Nielsen, B. Pallesen and M. S. Jorgensen., Bioresource Hemp, Wolfsburg., Sept. (2000).
11. J. Wiener, V. Kovacic, P. Dejlova., AUTEX Research Journal. **3**, 20 (2003).
12. W. Xia, L. Jian-Zhang, F. Yongming and J. Xiao-Juan., For. Stud. China., **8,** 1(2006).
13. S. K. Sharma, S. K. Shukla and D. N. Vaid., Defence. Sci. J., **33**, 3(1983).

HEMICELLULOSE EXTRACTION OF MIXED SOUTHERN HARDWOOD WITH WATER: EFFECT OF TEMPERATURE

Mehmet Sefik Tunc and Adriaan R.P. van Heiningen
The University of Maine, Orono, ME, 04469, USA

Abstract

Hemicelluloses derived from biomass are presently underutilized. In order to develop more profitable biorefinery processes, the mechanism responsible for hemicellulose removal by pretreatments has to be further explored. The hydrothermal dissolution profile of the wood components cellulose, hemicelluloses and lignin of a hardwood mixture during autohydrolysis in a modified Dionex ASE-100 (Accelerated Solvent Extraction) operating at high pressure (around 150 atm.) is described. A southern US hardwood chip mixture was subjected to pre-extraction with water from 130 to 170 °C for 100 minutes. The amount of cellulose, hemicelluloses, lignin, acetyl groups and 4-O-methyl-glucuronic anhydride (4-O-MGA) remaining in the extracted wood were determined. The extract was analyzed for oligosaccharides and monosaccharides, lignin, acetic acid, 4-OMGA, furfural and hydroxymethylfurfural (HMF).

Introduction

Hemicelluloses are the second major sugar component in lignocellulosic materials. They are being considered for production of value-added products; ethanol, succinnic acid and other chemicals but their complex character and linkage to lignin presents a major hurdle. Presently the hemicelluloses are underutilized in the pulp and paper industry because they are degraded during Kraft pulping, the major chemical pulping process. Because the value obtained from hemicelluloses in the Kraft process is mostly limited to recovery of the low heating value by combustion, a more economical use would be to extract them by (suitable) pretreatment methods, followed by conversion to higher value-added products [1]. Interest in isolation of hemicelluloses from biomass has greatly increased in recent years [2-9]. Several methods such as steam explosion, organic solvents, alkali, dilute acid, enzyme treatment and autohydrolysis have been suggested for removal of hemicelluloses from biomass. Because water is the only reagent in autohydrolysis, it is an environmentally friendly and inexpensive process compared to dilute mineral acid prehydrolysis [10].

Selective extraction of hemicelluloses from hardwoods can be accomplished with water at optimal operation conditions (temperature and time) [3,5,11]. Hemicellulose extracts (liquor) after autohydrolysis of hardwoods mainly contain oligosaccacharides (mainly xylooligosaccharide), monosugars (mainly xylose and mannose), low molecular weight cellulose, lignin, acetic acid, uronic acid and dehydration / decomposition products such as furfural, HMF, levulinic acid and formic acid [3,4,5,11,12]. Xylose based oligosaccharides can be isolated from biomass (lignocellulosec materials) and they have potential use in pharmaceutical, food, agriculture and papermaking industries [13-20].

In order to investigate the possibility to gain xylan in oligomeric form in high yield from a Southern hardwood mixture (SHM), a series of hydrothermal pretreatments of SHM were carried out. In this paper, the influence of temperature on the dissolution of hemicellulose, degradation of cellulose and delignification of a hardwood mixture when treated with water will be discussed.

Materials and methods

Materials

The Southern hardwood mixture (SHM) chips were air dried, then ground in a Wiley mill and the fraction passing 2 mm holes was stored in double plastic bags for later experimental use. The wood extractives were removed from SHM with dichloromethane to eliminate interference during the analysis of carbohydrates and lignin in the extracted wood. The chemical composition of extractives-free SHM is summarized in table 1. The amount of cellulose and hemicellulose are 42.3 and 29.0 % respectively.

Table 1. The chemical composition of original extractives-free SHM

g/100 g o.d wood								
Lignin	**Ash**	**AcG**	**MGA**	**Ara**	**Gal**	**Glu**	**Xyl**	**Man**
28.6	0.4	3.3	4.5	0.5	1.0	43.7	15.5	2.2

AcG: Acetyl groups
MGA: 4-O-Methyl-Glucuronic-Anhyride

Pre-extraction (autohydrolysis) of wood

Hemicelluloses were pre-extracted with water during 100 minutes of extraction time at temperatures ranging from 130 to 170 °C in a modified Dionex ASE-100 extractor. The liquor to wood ratio (L/W) was approximately 3.7:1.

Analysis of wood particles

The moisture content of the milled wood particles was determined by drying a representative sample at 100 ±5 ºC in a convection oven overnight. The ash content was determined according to TAPPI standard method T211 om-85. The acid insoluble lignin content, Klason lignin, was determined according to a method by Effland [21], while the acid soluble lignin content was determined by UV-Vis spectroscopy. The uronic anhydride content was determined using the chromophoric group analysis method developed by Scott [22]. The mono sugar content of the hydrolysate which was produced by double hydrolysis with sulfuric acid [23] was determined by High Performance Anion Exchange Chromatography with Pulse Amperometric Detection (HPAEC-PAD). Acetic acid in the hydrolysate was determined by HPLC using a refractive index detector and BIO-RAD Aminex HPX-87H column.

Analysis of solid and liquid phases after pre-extraction

The solid residue after pre-extracting was washed with 1.5 reactor volumes of water, air dried and subjected to the same analysis as applied to the original wood. The extraction yield was calculated from the difference in oven dry wood weight before and after pre-extraction. The pH of the liquid phase, called hemicellulose extract, after pre-extraction was recorded. The hemicellulose extract was analyzed for monosugar content in the HPAEC-PAD by direct injection and after one hour hydrolysis with 4% H_2SO_4 at 121 °C in an autoclave. The yield of oligomers was calculated from the increase in monosugar anhydride content due to acid hydrolysis. Furfural, HMF and acetic acid in the hemicellulose extract were determined using the HPLC system described earlier.

Results and discussion

Retention on wood

The retention of the major wood components; cellulose, hemicellulose and lignin as well as wood acids; acetyl groups and 4-OMGA, linked to hemicelluloses; during autohydrolysis are shown in figure 1. It is obvious that cellulose stayed intact during dissolution due to its high molecular weight and crystalline nature, while a small amount of delignification took place. The same behavior was observed by others [4,5]. Lignin removal only became significant at more severe conditions (T>160 °C). The majority of wood dissolution was due to hemicellulose removal. The retentions of acetyl groups and uronic anhydride bound to hemicelluloses also decreased dramatically with increasing temperature (see figure 1).

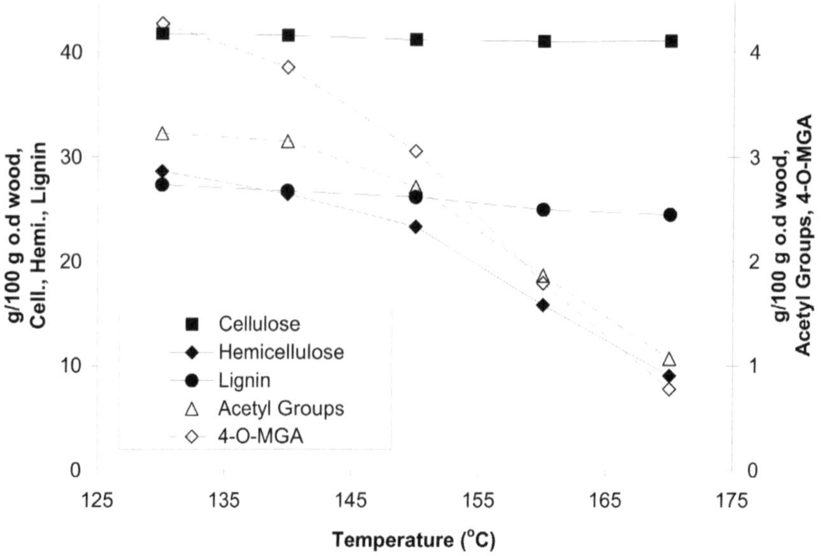

Figure 1. Cellulose, hemicellulose, lignin, acetyl groups and 4-O-MGA retentions of SHM versus temperature

Acetyl groups and 4-OMGA are generally associated with xylan in hardwood [24] and the molar ratio of acetyl groups and 4-OMGA to xylose remaining in extracted wood are shown in table 2. The molar ratio of acetyl groups to xylose in the extracted wood remained as high as in the original wood regardless of extraction temperature. Thus, xylan retained in the extracted wood remained acetylated during autohydrolysis regardless of temperature. This suggests that acetyl groups dissolved along with xylan. The ratio of 4-OMGA to xylose in the extracted wood was almost constant at low temperatures (T<150). However, the ratio decreased above 150°C. This indicates that 4-OMGA dissolved with xylan at low temperatures while 4-OMGA was preferentially removed at high temperatures (T>150 °C). This is surprising because it is well known that the glycosidic bond between methylglucuronic acid and xylan is more stable than the glycosidic bond between the acetyl groups and the xylan polymer [25]. The explanation may be that the uronic acid groups were still bound to monomeric or oligomeric xylosaccharides in the extract, while a more linear xylan remained in the wood.

Table 2. Molar ratios of acetyl groups and 4-O-MGA to 10 moles of xylose in wood

	Per 10 mole xylose (remaining in wood)	
Temp.(°C)	**Acetyl Groups**	**4-OMGA**
Wood	**6.5**	**2.2**
130	**6.5**	**2.2**
140	**6.5**	**2.0**
150	**6.1**	**1.7**
160	**6.3**	**1.5**
170	**6.5**	**1.2**

Composition of hemicellulose extract

The sugar content of the hemicellulose extract is summarized in table 3. It is apparent that xylooligosaccharides were the most abundant component in the extract. A similar observation was reported for dissolution of Barley Husks by Garrote *et al.* [26]. It is obvious from table 3 that mannan, glucose and xylan in the hemicellulose extract mainly existed in oligomeric form, while arabinan and galactan present were in both monomeric and oligomeric forms. Noticeable amounts of furfural and HMF were only generated at high temperatures. The amounts of furfural and HMF generated at 170 °C were approximately 0.33 and 0.03% based on original wood weight respectively.

Table 3. Composition of hemicellulose extract from liquid phase

	Monomeric Anhydro-monosugars (g/100 g o.d wood)					Oligomeric Anhydro-monosugars (g/100 g o.d wood)				
Temp.(°C)	Ara	Gal	Glu	Xyl	Man	Ara	Gal	Glu	Xyl	Man
130	0.02	0.00	0.01	0.00	0.00	0.11	0.10	0.42	0.08	0.09
140	0.07	0.02	0.01	0.01	0.00	0.16	0.17	0.85	0.36	0.14
150	0.14	0.04	0.03	0.04	0.00	0.22	0.29	1.03	1.82	0.27
160	0.22	0.10	0.05	0.23	0.01	0.16	0.44	1.14	5.83	0.51
170	0.18	0.18	0.13	0.95	0.06	0.08	0.47	1.16	8.23	0.77

Since the amount of extracted oligomeric xylan was much higher than that of oligomeric mannan (see table 3), it is assumed that all acetyl groups and 4-OMGA in the hemicellulose extract were associated with xylooligosaccharides. Based on this assumption the molar ratios of acetyl groups and 4-OMGA bound to the dissolved xylooligosaccharides were calculated, and the results are summarized in table 4. It is apparent that at mild extraction conditions (T<150 °C), the extracted oligosaccharides were not deacetylated since the molar ratio of acetyl groups and 10 moles of xylose in the original wood was 6.5. However, deacetylation of extracted oligosaccharides took place at temperatures higher than 150 °C, and leaded to a minimum acetyl group to xylose ratio at around 160 °C. The ratio of 4-OMGA linked to xylooligosaccharide to xylose in hemicellulose extract was almost constant within experimental error during dissolution regardless of temperature (see table 4).

Table 4. Molar ratio of Acetyl groups and 4-OMGA to 10 mole xylose (as oligosaccharides) in hemicellulose extract

	Per 10 mole xylose (oligomeric form)	
Temp.(°C)	Acetyl Groups	4-OMGA
Wood	6.5	2.2
130	-	2.7
140	6.8	2.5
150	4.0	2.3
160	3.4	1.7
170	5.0	2.5

Dissolution of main hemicellulose, xylan

The dissolution profile of xylan polymer versus temperature is depicted in figure 2. It is obvious that xylan dissolved mainly in oligomeric form during the autohydrolysis (pre-extraction with water). Xylan dissolved mostly as oligosaccharides at low temperatures (T<150 °C) and as temperature increased (reaction conditions become more severe) the oligosaccharides depolymerized to monomeric xylose by acid hydrolysis [10]. Furfural started to be formed at high temperatures (T> 150 °C) and its generation increased with increasing temperature (see figure 2).

The amount of the xylan missing in the hemicellulose extract during pre-extraction was calculated from the xylan mass balance. Missing xylan in figure 2 most likely represented the portion of xylan converted to decomposition products (DP) and as can be seen in figure 2 it increased with increasing temperature.

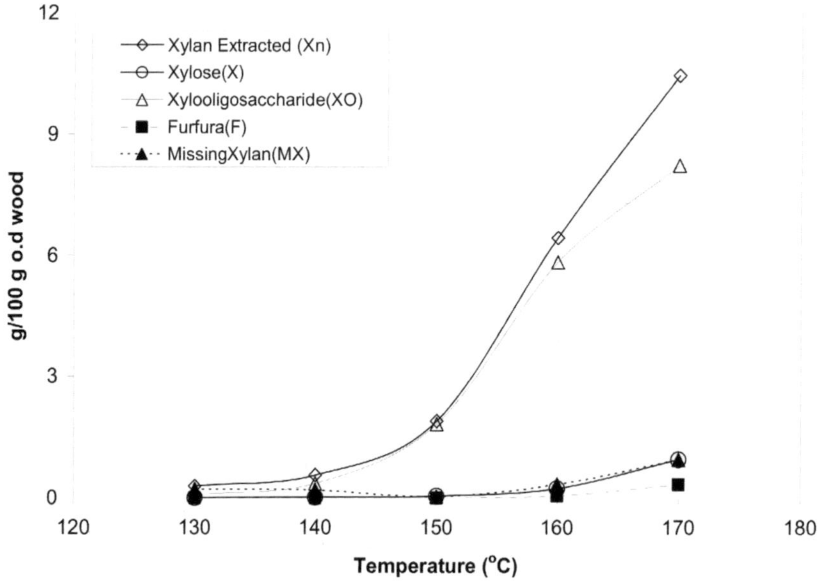

Figure 2. Xylan dissolution of SHM with water

The extraction yields of the monomeric and oligomeric forms of xylan and glucomannan hemicelluloses are plotted against pH in figure 3. It is clear that the hemicellulose removal increased rapidly at pH values below 4.3.

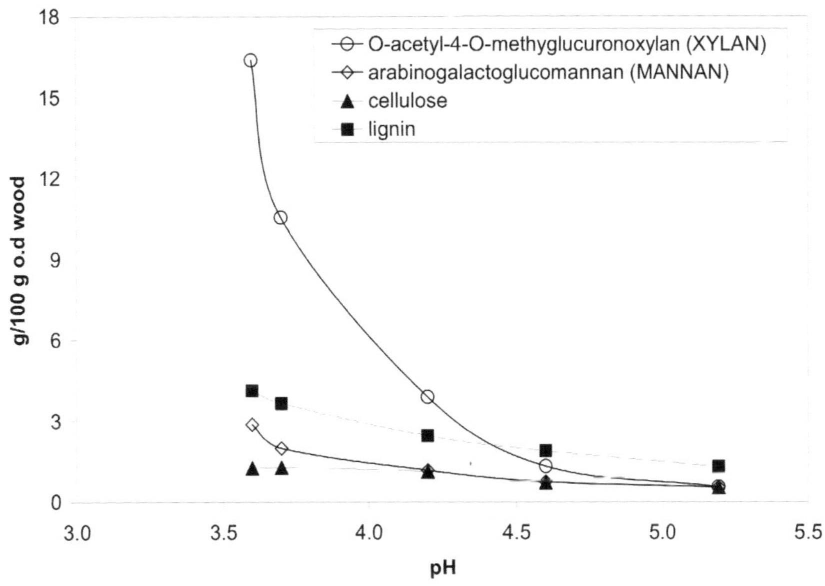

Figure 3. Dissolution of major hemicelluloses, cellulose and lignin of SHM versus pH

Conclusions

Presently the economic value generated from hemicelluloses in biomass is mostly derived from their heating value which is only half of that of lignin. However hemicelluloses may be used as a new feed stock for higher value added products, generating new income within an integrated biorefinery. The present study showed that xylooligosacchardies could be isolated from hardwood in high yield without extensive delignification and degradation of celluloses. Approximately 70% of hemicelluloses in original wood extracted at 170 °C. The retention of wood acids, acetic and uronic was decreased dramatically with increasing temperature. Xylan was the most abundant species in the reaction medium, especially at temperatures higher than 150 °C. Xylan mainly dissolved in oligomeric form, except at 170 °C when significant hydrolysis to xylose took place. Xylan retained in the extracted wood was highly acetylated regardless of temperature. The molar ratio of 4-OMGA linked to xylan polymer to xylose in the extracted wood remained the same as that in the original wood at low temperatures (T<150 C), while this ratio decreased at extraction temperatures above 150°C. The ratio of 4-OMGA to xylose in xylooligosaccharide in the extract was the same as that in original wood during dissolution regardless of temperature. The dehydration and decomposition reactions of dissolved carbohydrates took place at high temperatures (T>160 °C). The final pH of the hemicellulose extracts after 100 minutes of autohydrolysis decreased from 5.2 to 3.7 over the temperature range of 130- 170°C due to wood acids released. The extraction of the hemicelluloses increased rapidly when the extract pH was below 4.3.

References

1. van Heiningen, A.R.P. (2006), "Converting a kraft pulp mill into an integrated forest biorefinery," Pulp and Paper Canada, 107(6), pp. 38 - 43

2. Koshisima, T., Timell, T. E. and Zinbo, M. (1965), "The Number-Average Molecular Weight of Native Hardwood Xylans," Journal of Polymer Science., Part C, 11, pp. 265-279.

3. Casebier, R.L., Hamilton, J.K. and Hergert, H.L. (1969), "Chemistry and mechanism of water prehydrolysis on southern pine wood," Tappi, 52(12), pp. 2369-2377.

4. Casebier, R.L., Hamilton, J.K. and Hergert, H.L. (1973), "Chemistry and mechanism of water prehydrolysis on black gumwood, effect of temperature at constant time," Tappi, 56(3), pp. 135-139.

5. Garrote, G., Dominguez, H. and Parajo, J.C. (1999), "Mild autohydrolysis: an environmentally friendly technology for xylooligosaccharide production from wood," Journal of Chemical Technology and Biotechnology, 74, pp. 1101-1109.

6. Fang, J.M., Sun, R.C. and Tomkinson, J. (2000), "Isolation and characterization of hemicelluloses and cellulose from rye straw by alkaline peroxide extraction," Cellulose, 7, pp. 87-107.

7. Glasser, W.G., Kaar, W.E., Jain, R.K., Sealey, J. E. (2000) "Isolation options for non-cellulosic heteropolysaccharides (HetPS)," Cellulose, 7, 299-317.

8. Puls, J. and Saake, B. (2004), "Industrial Isolated Hemicelluloses," in: Gatenholm, P., Tenkanen, M. (Eds.), Hemicelluloses: Science and Technology. ACS Symp. Series 864, ACS. Washington, DC pp. 24-37.

9. Stalbrand, H., Lundqvist, J., Andersson, A., Hagglund, P., Anderson, L., Tjerneld, F., Jacobs, A., Teleman, A., Dahlman, O., Palm., M. and Zacchi., G. (2004), "Isolation, Characterization, and Enzymatic Hydrolysis of Acetyl-Galactoglucomannan from Spruce (Picea abies)," in: Gatenholm, P., Tenkanen, M. (Eds), Hemicelluloses: Science and Technology. ACS Symp. Series 864, ACS, Washington, DC, p. 66-78.

10. Conner, A.H. and Lorenz, L.F. (1986), "Kinetic modeling of hardwood prehydrolysis. Part III: Water and dilute acetic acid prehydrolysis of southern red oak," Wood Fiber Sci., 18, pp. 248-263.

11. Tunc, M.S. and vanHeiningen, A.R.P. (2007), "Hemicellulose Pre-extraction of Southern Hardwood Mixture with Water," *14th Int. Symp. on Wood and Pulping Chemistry,* Durban, South Africa, June 25-28 (2007) Oral Proceedings

12. Lora, J.H. and Wayman, M. (1978), "Delignification of hardwood by autohydrolysis and extraction," Tappi, 61(6), pp. 47–50.

13. Jaskari, J.; Kontula, P.; Siitonen, A.; Jousimies-Somer, H.; Mattila-Sandholm, T. and Poutanen, K. (1998), "Oat β-glucan and xylan hydrolysates as selective substrates for Bifidobacterium and Lactobacillus strains," Appl. Microbiol. Biotechnol., 49, pp. 175.

14. Okazaki, M.; Fujikawa, S. and Matsumoto, N. (1990), "Effect of xylooligosaccharide on the growth of Bifidobacteria," Bifidobacteria Microflora, 9, pp. 77.

15. Okazaki, M.; Koda, H.; Izumi, R.; Fujikawa, S. and Matsumoto, N. (1991), "Effect of xylooligosaccharide on the growth of intestinal bacteria and putrefaction products," J. Jpn. Soc. Nutr. Food Sci., 44, pp. 41.

16. Dohnalek, M. I. H.; Ostrom, K. M. and Hilty, M. D. (1998), "Use of indigestible oligosaccharides to prevent gastrointestinal infections and reduce duration of diarrhea in humans," U.S. Patent 5827526.

17. Campbell, J. M.; Fahey, J.C. and Wolf, B.W. (1997), "Selected indigestible oligosaccharides affect large bowel mass, cecal and fecal short-chain fatty acids, pH and microflora in rats," J. Nutr, 127, pp. 130.

18. Stone, A. L.; Melton; D. J. and Lewis, M. S. (1998), "Structure-function relations of heparin-mimetic sulfated xylan oligosaccharides: inhabitation of Human Immunodeficiency Virus-1 infectivity in vitro," Glycoconjugate J., 15, pp. 697-712.

19. Ebringerova, A. and Heinze, T. (2000), "Xylan and xylan derivatives-biopolymers with valuable. 1. Naturally occurring xylan structures, isolation procedures and properties." Macromol. Rapid Commun, 21, pp. 542-556.

20. Kayserilioglu, B. S.; Bakir, U.; Yilmaz, L.; Akkas, N. (2003), "Use of xylan, an agricultural by-product, in wheat gluten based biodegradable films: mechanical, solubility and water vapor transfer rate properties," Bioresource Technology, 87, pp. 239-246.

21. Effland, M.J. (1977), "Modified procedure to determine acid-insoluble lignin in wood and pulp," Tappi, 60(10), pp. 143-144.

22. Scott, R.W. (1979), "Colorimetric determination of hexuronic acids in plant materials," Analytical Chemistry, 51(7), pp. 936-941.

23. Davis, M.W. (1998), "A Rapid Modified Method for Compositional Carbohydrate Analysis of Lignocellulosics by High pH Anion-Exchange Chromatography with Pulsed Amperometric Detection (HPAEC/PAD)," J. Wood Chemistry and Technology, 18(2), pp. 235-252.

24. Sjostrom, E. (1993). Wood Chemistry Fundamentals and Applications, 2nd Edition, Academy Press Inc.

25. Harris, J.F. (1975), "Acid hydrolysis and dehydration reactions for utilizing plant arbohydrates," Appl. Polym. Smp. 28, pp. 131-144.

26. Garrote, G., Dominguez, H. and Parajo, J.C. (2004), "Production of Substituted Oligosaccharides by Hydrolytic Processing of Barkley Husks," Ind. Eng. Chem. Res., 43(7), pp. 1608-1614.

Property of Solid Residue in Woody Biomass Saccharization Process by Hot-Compressed Water Method

Ayumu HIRAKAWA*1, Nobusuke KOBAYASHI*1, Shigenobu HATANO*1, Yoshinori ITAYA*1, Yasuhiro TANABE*1, Jun KOBAYASHI*2, Shigekatsu MORI*1

[1]Department of Chemical Engineering, Nagoya University, Nagoya, Japan
[2]National Institute for Environmental Studies

INTRODUCTION

Producing bio-ethanol from woody biomass has attracted a big attention in terms of alternative clean energy. Hydrolysis of woody biomass was the important process for producing bio-ethanol, and then the efficient and economical hydrolysis process of woody biomass has been studied in Japan.

Enzymatic saccharization process was one of the hydrolysis methods, which was able to achieve the high yield of monosaccharide from cellulose, but the enzymatic saccharization process usually required the pretreatment process for the woody biomass, since enzyme was usually not hydrolyze hemi-cellulose and lignin. The high enzyme price and enzyme recovery was also a problem for the commercialization. Therefore, there are lots of researches regarding the enzyme development and pretreatment processes, which can hydrolyze the complicated woody biomass efficiently.

For the efficient enzyme saccharization, it was very important to remove the hemi-cellulose or lignin from woody biomass before enzymatic treatment, since accessibility of the enzyme to the cellulose was the key of efficient saccharization. There are several researches for the development of enzyme accessibility. One of the treatments called hot-compressed water process (HCWP) was recently studied for the woody biomass saccharization. HCWP was the method which can hydrolyze wood component to the saccharine in the hot-compressed water without any acid. Hemi-cellulose component in woody biomass was usually decomposed completely in HCWP, but the yield of saccharine from cellulose was low in the mild reaction condition. The combined process of HCWP and enzymatic saccharization process was recently developed, since HCWP was good for the hydrolysis of hemi-cellulose and enzyme treatment was good for the hydrolysis of cellulose. But, the optimum combination process of HCWP and enzyme saccharization was not understood so far. Especially, the relationship between the characteristics of solid products from HCWP and enzyme saccharization was understood at all.

In this paper, the physical characteristics of solid residue (SR) in HCWP were studied for the efficient enzyme saccharization process. As the first step, particle size, pore volume, specific surface area, cellulose crystallinity before and after the HCWP were measured, and the decompose mechanism of wood component in HCWP was discussed.

EXPERIMENTAL

Experimental Apparatus and Procedure

1. Analysis of the hydrolysis decomposition product

Figure 1 shows the schematic diagram of microwave heating device. Before the liquefaction experiment, woody biomass slurry was prepared by using the wood powder. Cedar chips were pulverized by the vibration mill, and the pulverized powder was screened to under 60 μm. The woody slurry of 60 g was made by mixing 6 g pulverized wood and 54 g distilled water together. Well-mixed slurry was fed in the reactor and heated at a rate of 80 K/min up to the target temperature and kept for 40 min. After the reaction, the reactor was cooled by water. The reaction product was taken from the reactor, and it was separated solid residue from liquid product by the suction filtration. After the filtration, the liquid products were analyzed by HPLC.

2. Analysis of the solid residue

Before the liquefaction experiment, woody biomass slurry was prepared by using the wood powder as described below. Cedar chips were pulverized by the vibration mill and the powder was screened to 20-100 μm, 100-212 μm, 212-500 μm. The slurry of 500 g was made by mixing 40 g wood powder and 360 g distilled water together.

Figure 2 shows the schematic diagram of the hydration experiment unit. This experiment unit was the batch type reactor (autocrave), and it was consists of a pressure vessel (reactor), heater, pressure gauge, agitator for mixing wood slurry, thermocouple, temperature controller and rpm controller. Well-mixed wood slurry was fed in the reactor. Nitrogen gas was purged for 15 minutes with agitating the wood slurry to create an inert condition. After replacing the air, the reactor was heated at a rate of 12 K/min up to the target temperature of 453 K and kept for 10-40 min. After the reaction, the reactor was cooled by ice water immediately to prevent the secondary reaction. The reaction product was taken from the reactor, and it was separated solid residue from liquid product by the suction filtration. After the filtration, solid residue was washed by distilled water several times. The solid residue was dried at 378 K for 24 hours in an oven. A part of the solid residue was pulverized more by a planetary ball mill for 60 min. The measurement of particle size distribution, cumulative pore volume distribution, surface area, and crystallinity were carried out. The particle size was measured by laser-diffraction particle diameter analyzer, pore size and surface area was measured by AUTOSORB, and XRD peak pattern was measured for analyzing cellulose crystallinity.

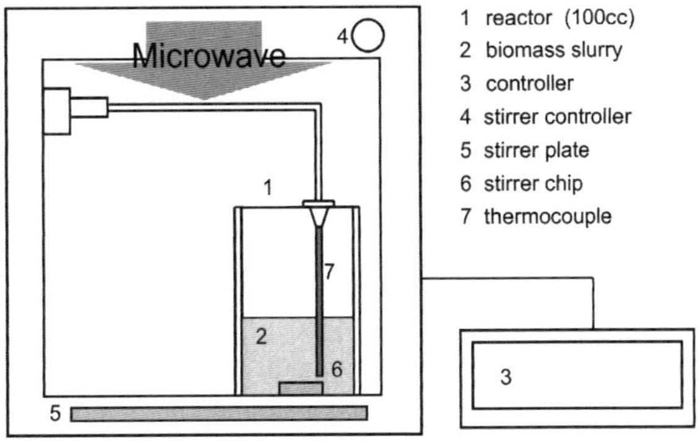

Figure 1. Schematic diagram of microwave heating device

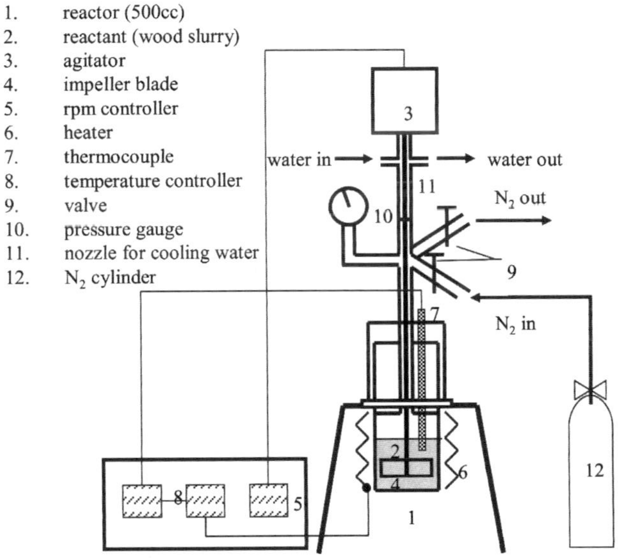

Figure 2. Schematic diagram of autocrave

Experimental Results and Discussion

1. Analysis of the hydrolysis decomposition product

Figure 3 shows the HPLC chromatogram of the products from woody biomass decomposition. The peaks of oligosaccharides, monosaccharide and decomposition products from monosaccharide were observed at the different operation temperature of 433 to 473 K. The result indicates that lower-molecular-weight products decomposed easily with increasing the reaction temperature. A large peak of monosaccharide was observed at the operation temperature 453 K and 473 K. Decomposition products of monosaccharide at 473 K were less than that at 453 K, and almost no products were decomposed at 453 K. Therefore, 453 K is the optimum operation temperature to yield monosaccharide from hemi-cellulose as the pretreatment process for the enzyme saccharization.

Figure 4 shows the liquefaction yield of total decomposition products from woody biomass at 453 K. The yield of decomposition products was calculated by **Eq.1**.

Liquefaction yield [%]

$$= 1\text{-(solid residue weight after HCWP(dry)[g])/(initial wood weight(dry)[g])} \times 100 \quad \textbf{(Eq.1)}$$

The liquefaction yield increased with increasing the reaction time. The liquefaction yield saturated after the reaction time of 40 min. Considering HPLC result and liquefaction yield result together, almost all hemi-cellulose component was decomposed at the reaction time of 40 min at 453 K.

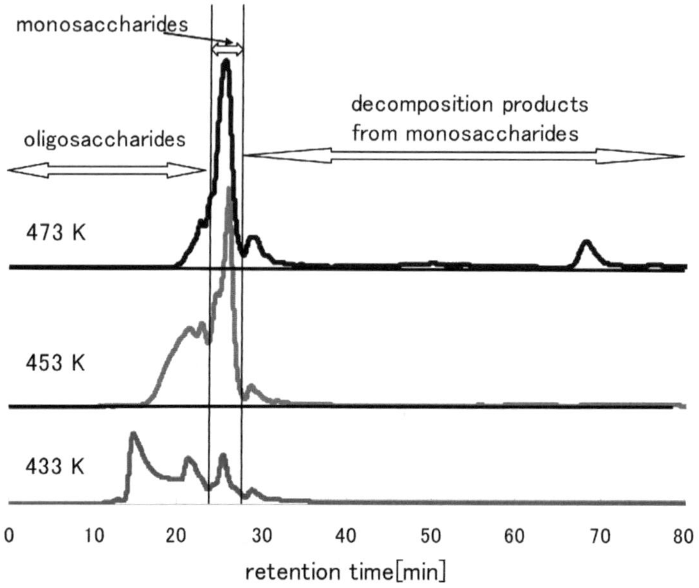

Figure 3. HPLC chromatogram of the liquid products at the different operation temperature

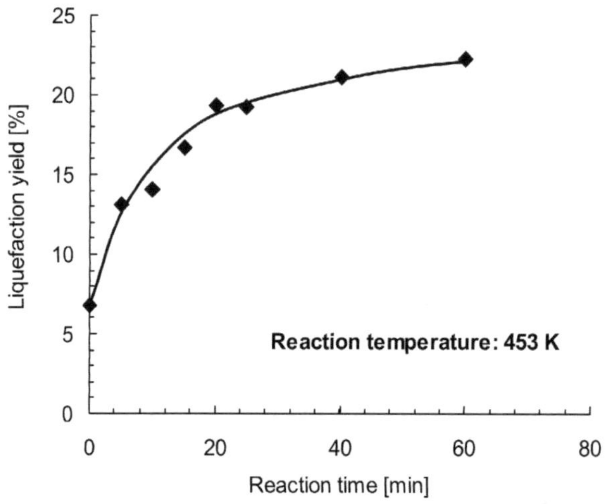

Figure 4. Liquefaction yield at the operation temperature 453 K

2. Analysis of the solid residue

Particle size distribution

Figure 5 and **Figure 6** show the particle size distribution of solid residue after the hydrolysis treatment. Raw woody powder screened to 20-100 μm and 212-500 μm (SR20-100, SR212-500) was used. The particle diameter decreased with increasing the reaction time. The decrease rate was larger at the reaction time 0-20 min than 20-40 min due to the liquefaction yield difference (**Fig.4**). The result indicates that the decrease of particle diameter with hydrolysis is due to the decompose of hemi-cellulose from the surface of wood powder gradually. The particle size of solid residue pulverized by a planetary ball mill after hydrolysis (SRP) was drastically decreased.

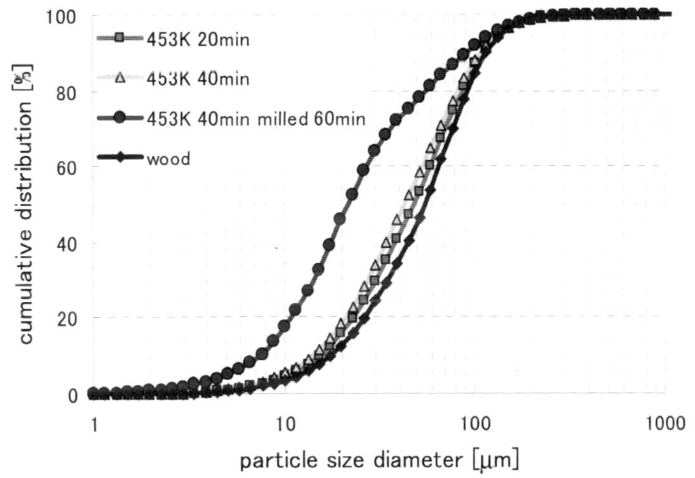

Figure 5. Particle size distribution of solid residue hydrolyzed from wood powder screened to 20-100 μm

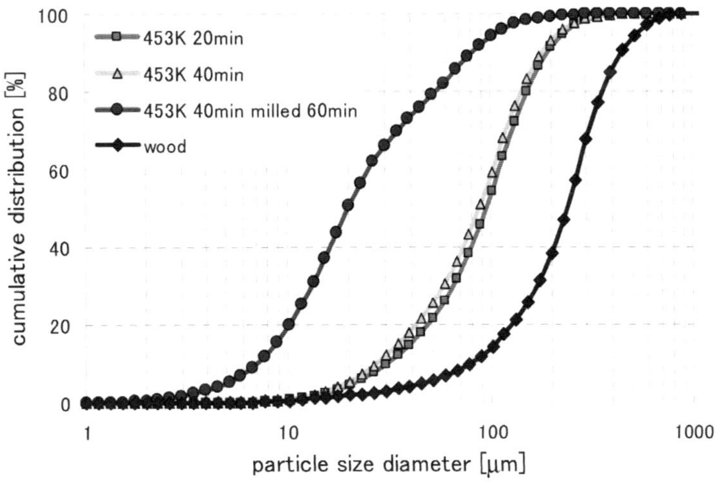

Figure 6. Particle size distribution of solid residue hydrolyzed from wood powder screened to 20-100 μm

Pore size distribution

Figure 7 and **Figure 8** show the cumulative pore volume distribution of SR20-100, SR212-500, respectively. The pore volume of SR increased than the wood. This result indicates that the decompose reaction of hemi-cellulose not only progressed from the surface, but also developed from the inner particle pore. The cumulative pore volume of SRP212-500 was increased than SR212-500, but the SRP20-100 was decreased than SR20-100.

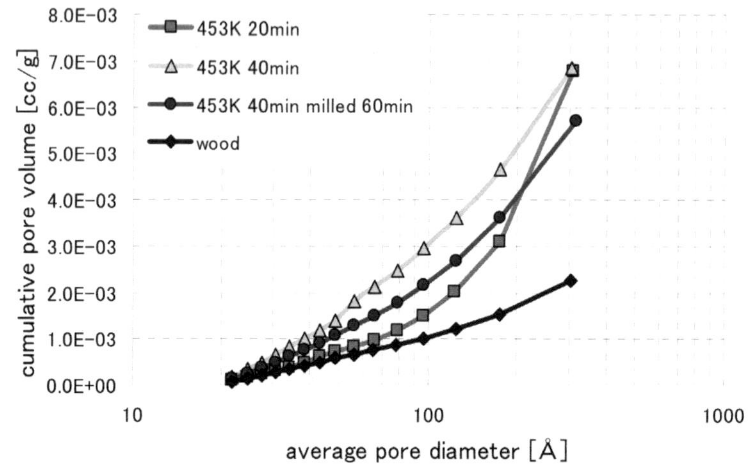

Figure 7. Cumulative pore volume distribution of solid residue hydrolyzed from wood powder screened to 20-100 μm

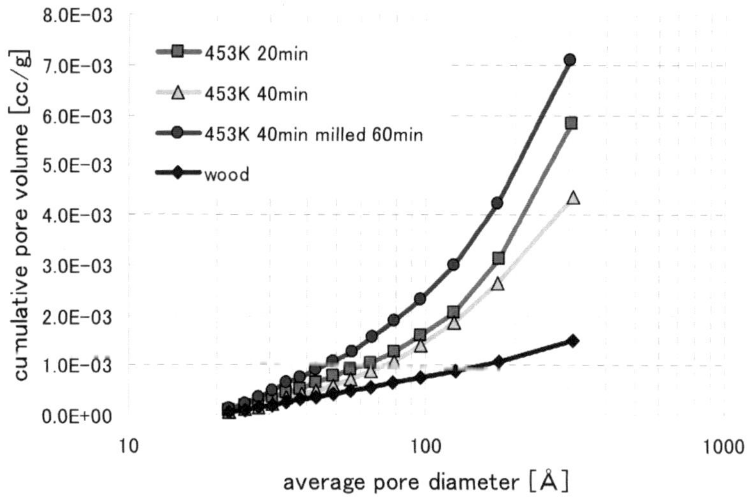

Figure 8. Cumulative pore volume distribution of solid residue hydrolyzed from wood powder screened to 212-500 μm

Specific surface area

Figure 9 shows the specific surface area of solid residue. The surface area increased with the reaction time of the liquefaction due to decrease of the particle size and increase of the pore volume. The adding pulverization increased the surface area of SRP212-500, but the surface area of SRP20-100 was decreased than SR20-100, because the pore volume was decreased with the milling.

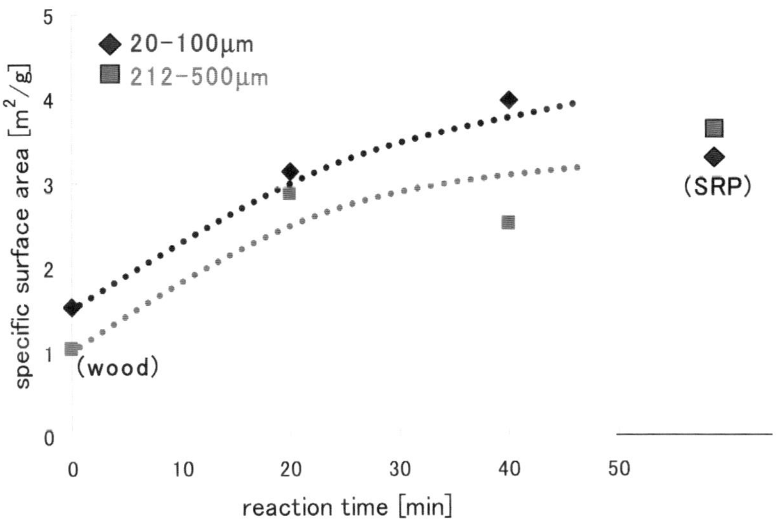

Figure 9. Specific surface area of solid residue

XRD peak pattern of solid residue

Figure10 shows the XRD peaks of SR and SRP. The peaks observed at $2\theta = 16^\circ$ and 23° was identical peak of cellulose crystal. The peak intensity-16, 23° of SR was no difference with the peak intensity of wood. This result indicates that the cellulose crystal was not decomposed with hydrolysis at 453 K. However, the peak intensity-16, 23° of SRP was smaller than wood because the cellulose crystal was destructed with the milling. The broad peak at $2\theta = 18^\circ$ indicates the presence of the amorphous cellulose , hemi-cellulose, or lignin. The decrease of peak intensity-18° of SR was due to the decompose and dissolution of hemi-cellulose, and the increase of peak intensity-18° of SRP was due to the increase of the amorphous cellulose.

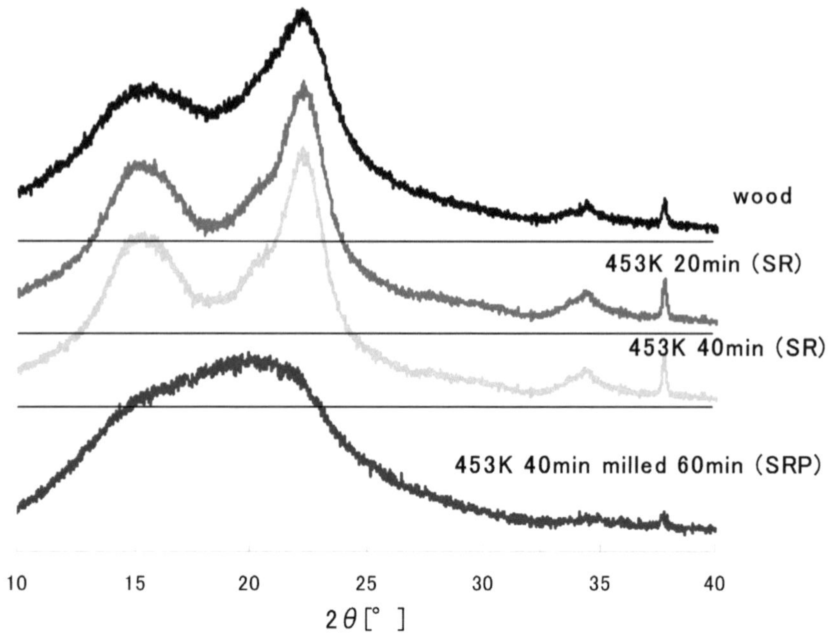

Figure 10. XRD peak pattern of solid residue

CONCLUSIONS

The physical characteristics of solid residue in HCWP and solid residue pulverized after hydrolysis were measured and discussed in this paper. The following conclusions are summarized as;

(1) Hemi-cellulose component was efficiently decomposed to monosaccharide and almost no decomposition products from monosaccharide were observed at the operation temperature of 453 K in HCWP.

(2) Hemi-cellulose decompose reaction not only occurred from the particle surface, but also developed from the inner particle pore. The specific surface area was increased with the hydrolysis.

(3) The cellulose crystal was not decomposed at the operation temperature 453 K in HCWP.

(4) The pore volume and the surface area of the solid residue pulverized after hydrolysis were increased and the cellulose crystal was destructed.

415c Membrane Separation: An Essential Component For A Wood-Based Biorefinery

Christopher D Wood, Thomas E. Amidon, and **Shijie Liu**. SUNY College of Environmental Science and Forestry, 1 Forestry Drive, Syracuse, NY 13210

Starting from a wood-based biorefinery, this paper discusses the separation and purification platform chemicals from the wood extracts. When wood is subject to hot-water extraction, extractives, hemicelluloses and other components leach out into the extraction liquor. Up to 25% woody biomass can be leached out while the wood chips still remain rigid and maintains traditional utilities. The wood extracts can be fractionated for platform chemicals such as monomeric sugars, acetic acid, and methanol. Membrane units can be employed to accomplish the separation and concentration tasks. It has been found that the permeate flow rate increase with increasing pressure drop and decreases with increasing extract concentration. Besides using membrane units as a fractionation device, they can be employed to concentrate the sugar / chemical stream as well. As compared with evaporation concentration, membrane systems can significantly improve the energy efficiency.

415d Improvement of Bio-Oil Stability

Rebecca K. Toghiani, Chemical Engineering, Dave C. Swalm School of Chemical Engineering, Mississippi State University, 330 Swalm, Mississippi State, MS 39762, Ananda Sekar, Dave C. Swalm School of Chemical Engineering, Dave C. Swalm School of Chemical Engineering, Mississippi State University, 330 Swalm, Mississippi State, MS 39762, and Phil Steele, Forest Products Department, Dave C. Swalm School of Chemical Engineering, Mississippi State University, 330 Swalm, Mississippi State, MS 39762.

Fast pyrolysis of biomass generates a product bio-oil containing a complex assortment of chemicals. Long-term stability of the bio-oil product is degraded by the presence of compounds that continue to undergo reaction during storage. An investigation of post-pyrolysis treatment of the bio-oil using supercritical water was conducted. Supercritical water has been reported in the literature as a means to decompose biomass. The effect of supercritical water (SCW) on the chemical product distribution in bio-oil was examined through a series of experiments. Critical process parameters were reaction time, temperature, pressure, as well as the bio-oil/water ratio. A base-line analysis of the bio-oil was performed. Comparison of pre- and post-treatment analysis will reveal those compounds which are significantly affected by treatment.

TECHNICAL EVALUATION OF A HARDWOOD BIOREFINERY USING THE "NEAR-NEUTRAL" HEMICELLULOSE PRE-EXTRACTION PROCESS

Authors: H. Mao, J. M. Genco, A. van Heiningen and H. Pendse
Address: University of Maine, Department of Chemical Engineering
Orono, Maine USA 04469

ABSTRACT

A technical analysis was performed for a forest biorefinery that produces ethanol and acetic acid from hardwood in addition to Kraft pulp. The new process involves hemicellulose extraction using green liquor prior to conventional Kraft pulping. In the design reported here an existing impregnation vessel was converted into an extraction vessel prior to continuous pulping. The extracted liquor undergoes evaporation, hydrolysis, separation, fermentation and distillation for the production of acetic acid and ethanol. A computer model was developed for the hemicellulose extraction process using WinGEMS and ASPEN Plus software. The capital and operating cost for the process were estimated by standard methods. The amounts of ethanol and acetic acid produced are relatively small because of the small amount of hemicelluloses extracted, dictated by pulp quality and pulp yield considerations. The rate of return on investment varies between 13 and 18 percent depending upon the size of the pulp mill (750 to 1500 tonne/day) for the case where the utilities and waste treatment facilities are sufficiently large to handle the additional requirements for the process. If the utilities and waste treatment system have to be upgraded, then the rate of return on investment decreases to between 6% to 11% again depending upon the size of the pulp mill (750 to 1500 tonne/day).

INTRODUCTION

Background

Interest in renewable resources as a source of energy and chemicals has increased because of potential worldwide shortages of petroleum and natural gas and growing concerns regarding the accumulation of greenhouse gases in the atmosphere. The concept of an Integrated Forest Products Biorefinery (IFBR) is being advanced by a number of investigators, who envision converting cellulose, hemicellulose and lignin from woody biomass, dedicated annual crops and municipal waste into bioenergy and basic chemicals [1, 2, 3, 4, 5, 6, 7, and 8]. The IFBR concept pursued at the University of Maine [9] seeks to integrate pulp production and a number of by-products derived from hemicellulose. In this process a portion of the hemicellulose polymers is extracted from wood prior to pulping and converted into ethanol and acetic acid while using the extracted wood chips to produce market Kraft pulp.

The main purpose of the work reported here is to present the results of a technical and economic analysis of the hemicellulose extraction process. In this analysis an attempt was made to determine the economy feasibility of building a commercial biorefinery that is co-located at an existing Kraft pulp mill and is fully integrated with the mill in terms of mass and energy. The objective of this work was to determine the economic and technical feasibility of the "Near-Neutral" hemicellulose pre-extraction process integrated within an existing hardwood Kraft mill.

The biorefinery concept investigated here involves co-locating the pre-extraction and ethanol conversion processes at an existing Kraft pulp mill which was considered as the base case. The existing

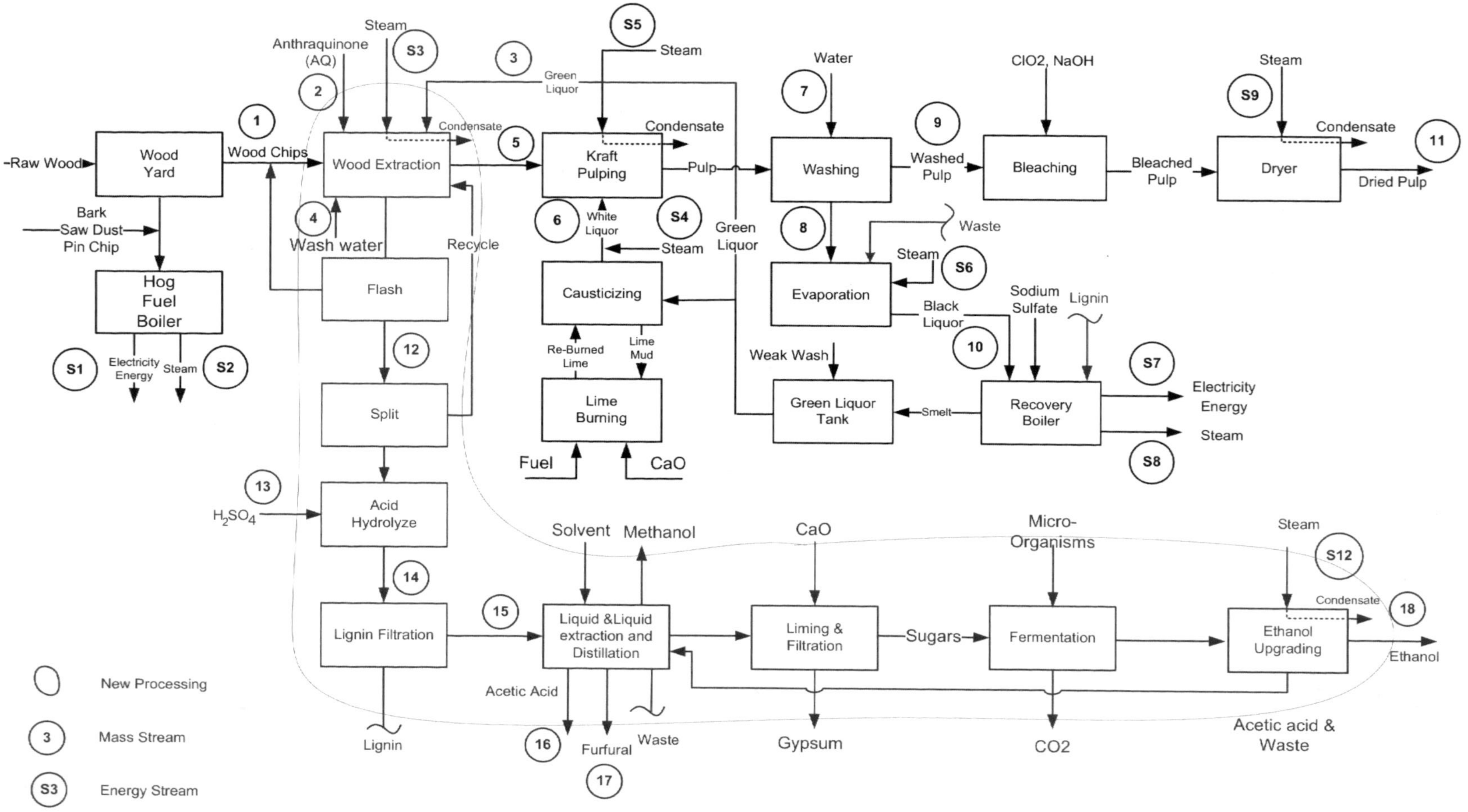

Figure 1. Modified Kraft Pulp Mill for the Production of Acetic Acid and Ethanol

pulp production was maintained constant and the hemicellulose extraction process was added to the fiber line. The feedstock to the biorefinery was assumed to be mixed US Northeast hardwoods (birch, beech, maple), while pulp production in the fiber line was being done using a continuous digester. In the analysis it was assumed that a separate impregnation vessel existed prior to the continuous digester that could be converted into the required extraction vessel.

The conceptual biorefinery process used in the design study reported here is shown in Figure 1. The patent pending extraction and modified cooking process [10] is carried out using green liquor (mostly Na_2CO_3 + Na_2S) and white liquor (mostly NaOH + Na_2S) respectively. The charge of green liquor used in the pre-extraction is about 3% (as Na_2O on dry wood) counted as total titratable alkali (TTA), while the charge of effective alkali (EA) in the modified kraft cook is reduced by 3% (as Na_2O on original dry wood) compared to conventional kraft cooking to the same bleachable grade hardwood pulp. A charge of 0.05% (on wood) of anthraquinone is included in the green liquor. The processing of the pre-extracted wood is shown in black and represents a conventional Kraft pulp mill (Figure 1). The unit processes required for hemicellulose extraction and conversion to ethanol and acetic acid are marked in blue. They include wood extraction for hemicellulose removal, flashing of the extract to produce preheating steam, recycling a portion of the extract back to the extraction vessel for the purpose of raising the solids content of the extract, acid hydrolysis using sulfuric acid for conversion of the oligomeric carbohydrates into mono sugars and cleavage of lignin-carbohydrate covalent bonds, filtration to remove precipitated lignin, liquid-liquid extraction followed by distillation to remove acetic acid and furfural from the sugar solution, liming to raise the pH to that required for fermentation, fermentation of five and six-carbon sugars and glucuronic acid to ethanol and finally distillation and upgrading the product to pure ethanol (99+%).

Process Advantages There are several important advantages associated with the "near-neutral" hemicellulose extraction process. Firstly, the quantity and quality of the pulp is unchanged compared to conventional Kraft cooking as shown in Figures 2 and 3 respectively. Figure 2 shows that pre-extraction with 3% green liquor (+0.05% AQ) at 160°C for 110 minutes followed by modified kraft cooking at 12% EA (30% sulfidity) gives the same yield (blue triangular symbols) based on original wood as regular Kraft cooking at 15% EA and 30% sulfidity (red star symbols). Figure 3 shows that the tear-tensile strength of the pre-extracted Kraft pulp is as good as or slightly better than that of the Kraft pulp produced using conventional methods.

A second advantage of the "near-neutral" extraction process is that the recovery cycle is off-loaded because the amount of organics in black liquor is reduced and less white liquor is needed for pulping. This change would allow a mill to significantly increase its pulp production rate if the recovery cycle is the bottleneck. Environmental advantages of the extraction process are that the methanol content of the black liquor is reduced by about 40%, while its TRS content is also diminished because of the lower white liquor charge and shorter pulping time during the modified Kraft cook. Finally, a new feed stock stream of sugars is produced which may be used for the production of biofuels and renewable chemicals.

PROCESS DESIGN

Wood Composition

The main constituents present in wood are cellulose (42.6%), hemicellulose (29.7%) and lignin (27.4%). Additionally there is about 0.3% inorganic (ash) and about 2 to 3% extractives which are low molecular weight compounds consisting of fats (acids and alcohols), waxes and sterols [11] in hardwoods.

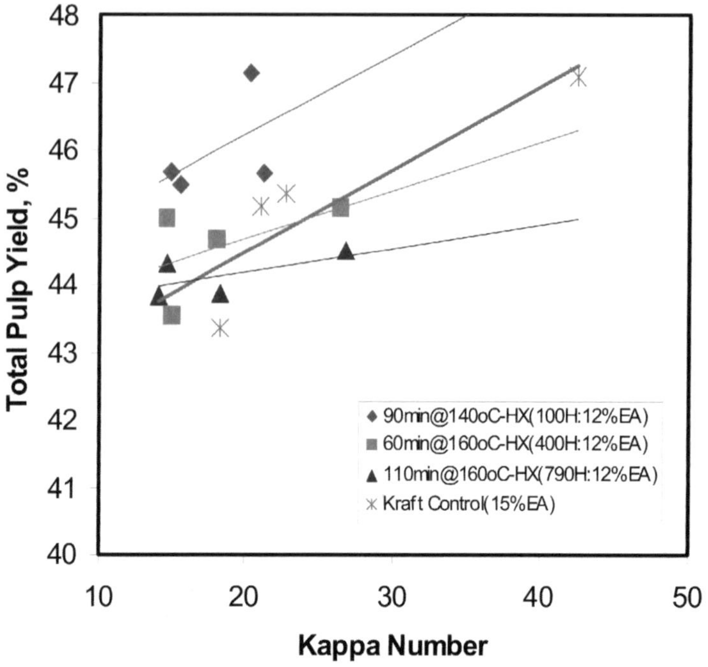

Figure 2 Yield vs. Kappa number

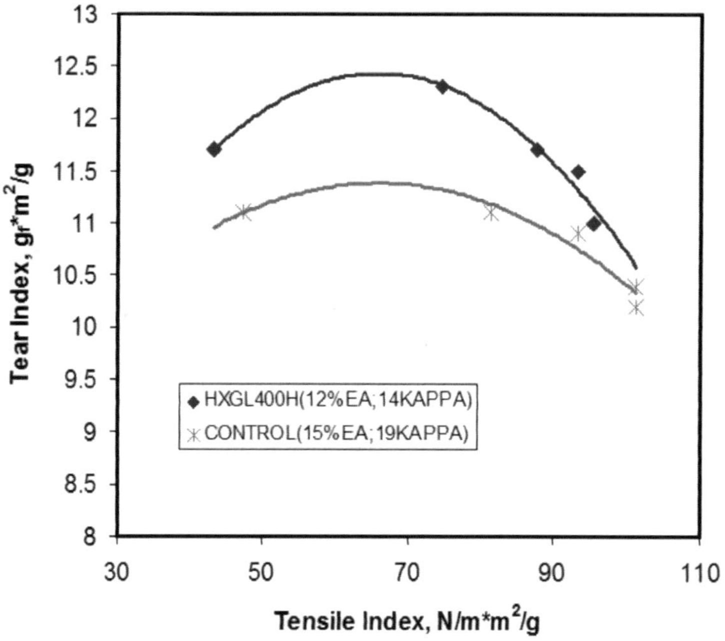

Figure 3 Tear Index vs. Tensile Index

Plant Sizes and Assumptions

The basis for the design study is summarized by Mao [12]. Three sizes of pulp mills were considered; a small plant producing 750 tonne/day of market Kraft pulp, a medium plant of 1,000 tonne/day and a large pulp mill of 1,500 tonne/day. The base case is an existing Kraft pulp mill and the technical economics were considered relative to the base case.

Pulping Conditions

The Kraft pulping conditions for the base case was 15% EA, 170 °C temperature to give an H-factor of 1500 hours. For the hemicellulose extraction process the pulping conditions the total H-factor was maintained the same; that is an H-factor of 700-hours was used in the extraction vessel and 800 hours in the cooking digester. Pulping conditions were 12% EA at 170 °C

Extraction Liquor

Several extraction liquors were considered; green liquor, sodium carbonate, sodium hydroxide, white liquor and pure water [12]. Green liquor was selected because it is a stream produced in the recovery cycle of the Kraft mill, the resulting near-neutral pH of the final extract avoids lignin precipitation, and the pulp yield (based on original wood) and physical properties of the pulp are maintained. The use of pure water as extraction liquid results in a low pH final extract and a lower overall pulp yield and lower pulp quality.

The composition of the extract used in the design analysis is given in Table 1 and is based on extraction experiments performed by van Heiningen [14]. The primary carbohydrates found in the extract are xylan and glucuronic anhydride which can be hydrolyzed and then fermented into ethanol and acetic acid. During the extraction process acetyl groups, which are side chains on the xylan hemicelluloses, are cleaved to give dissolved sodium acetate which will be converted to acetic acid in the hydrolysis stage of the plant. Some lignin is also extracted.

Process Simulation

The simulation model for the modified Kraft pulp mill first was developed primarily using the WinGEMS computer code [13] which was designed for solids/liquid systems occurring in the pulp and paper industry. Compound blocks, or subroutines designed to simulate complicated unit operations were used in this model. In performing the energy balance calculations distillation operations were modeled using ASPEN and by scaling data provided by Wooley et al. [7].

"Near-Neutral" Extraction Process

Wood is pre-steamed in a steaming vessel for 15 minutes and then heated to the extraction temperature of 160 °C in 60 minutes and then extracted for 110 minutes at 160 °C. Under these conditions approximately 10% of the mass of the wood is extracted (Table 1). The extraction liquid containing hemicellulose and lignin is separated from the chips for production of ethanol and acetic acid. The chips in the extraction vessel maintain their physical integrity during the extraction process except there is approximately a 10% loss of mass. After extraction the chips are conveyed into the pulping digester and heated to 170 °C where they are cooked for a sufficient time to achieve the desired H factor.

Lignin Removal and Acid Hydrolysis

After extraction, the extract liquor at high temperature (160 °C) and elevated pressure (110 psig) is sent to a flash tank to recovery steam and lower the temperature (130 °C) and pressure (20 psig). The dissolved solids content of the extract leaving the extraction vessel is approximately 8.5 %. The extract may then be further concentrated using a multiple effect evaporator system, but this was not considered in the present paper. The hydrolysis of the component carbohydrates is assumed to follow the kinetics specified by Garrote and co-workers [15] for hydrolysis of extract from Eucalyptus using sulfuric acid. The concentration range of sulfuric acid studied was 0.5 to 4 weight %. A high concentration of acid charge will reduce the hydrolysis time and lower the equipment cost, but it has high operation cost since the sulfuric acid which acts as a catalyst in the hydrolysis reaction must be recovered or neutralized.

Table 1
Composition of Extraction for
1,000 Tonne per day Pulp Mill Production

Composition of Extract	**Value**
Flow rate	3,171 Tonne/day
Suspension solids content	0%
Dissolved organics	5.61%
Arabinan	0.06%
Galactan	0.17%
Mannan	0.11%
Glucan	0.22%
Xylan	1.889%
Acetyl group	1.11%
4-O-MGA	1.33%
Lignin	0.728%
Dissolved inorganics	2.84%
H^+	0%
Na^+	1.11%
SO_4^{2-}	0.70%
HCO_3^-	1.03%
Water	2,903 tonne/day

In the design reported here, the hydrolysis conditions are 0.5% sulfuric acid, 130 C and a reaction time of 1.3 hours. At these conditions it was assumed that all the extracted lignin is precipitated and can be separated from the liquor by filtration. After lignin separation a proprietary liquid- liquid extraction process is used to extract acetic acid, furfural and traces of methanol from the mixture liquor. Based on this technology the recovery efficiency for acetic acid and furfural is about 90%. The extracted acetic acid and furfural are further purified by distillation and are sold as final products.

Liming and Gypsum Removal

After liquid- liquid extraction and acetic acid and furfural recovery, calcium oxide (CaO) is used to "lime" the extract liquor. The main function of liming is to adjust the pH to near neutral condition for ethanol fermentation. The lime also acts as a bactericide. The calcium precipitates the sulfate ions as gypsum, $CaSO_4{\cdot}2H_2O$, to a concentration level which can be tolerated by the fermentation culture. Gypsum is removed by filtration.

Ethanol and Acetic Acid Production

In the fermentation step [7, 16, and 17], it was assumed that C5 and C6 organics can be fermented simultaneously by using E.Coli (KO11). Also, glucuronic acid was assumed to be fermented into acetic acid and ethanol by using E. Coli B as described by Lawford and Rousseau [18]. The micro-organisms for conversion of the component sugars to ethanol would be produced on site. The efficiency of converting sugars and glucuronic acid into ethanol and acetic acid was set at 90% of the theoretical yield. Ethanol was processed after fermentation by pre-distillation to 50% purity and then further distillation to 95% which is the azeotrope concentration. The ethanol is then further concentrated to 99.9% by using molecular sieve technology [19].

Savings in the Lime Kiln

In the "near neutral" extraction process less white liquor is required in the cooking step. This will result in a corresponding decrease in the amount of calcium carbonate that needs to be removed in the white liquor clarifier and decomposed to lime in the kiln. This reduction in flow of $CaCO_3$ has a significant effect on the amount of energy required to operate the lime kiln. This is a distinct advantage of using green liquor rather than sodium hydroxide or white liquor in the pre-extraction step. The extraction step uses approximately 3% green liquor, calculated on dry wood basis and expressed at the total titratable alkali again expressed on a Na_2O basis. It allows a reduction in the effective alkali charge of white liquor in the modified Kraft cook of 3%; 15% EA in conventional cooking based upon dry wood basis and 12% EA in the modified cooking step. As a result, the reduction in $CaCO_3$ that must be processed in the lime cycle will be reduced by 80% based on the use of 30% Sulfidity in the white liquor or cooking liquor.

RESULTS

Production Rate

For a 1,000 tonne per day pulp mill the production rate for ethanol would be approximately 39.5 tonne/day (4.62 million gallons per year of 100% ethanol) and 46.6 tonne per day of acetic acid (3.97 million gallons per year of 100% acetic acid). In addition about 0.92 tonne per day of furfural are produced (320 tonne per year) and incinerated. Figure 4 illustrates the production rates for ethanol, acetic acid and furfural as a function of the production rate of the Kraft mill. Even for the 1,500 tonne per day Kraft mill, the amount of ethanol produced is relatively small when compared to a modern corn to ethanol plant which typically produces 50-100 million gallons per year [20]. Similarly, for comparison purposes, a modern acetic acid plant would have a production rate of 500,000 tonne per year, which is about 30 times larger than the capacity of the 1,000 tonne per day pulp mill [21]. However, although these wood derived chemical production rates are relatively small, the production of ethanol and acetic acid represents significant sources of revenue for the pulp mill.

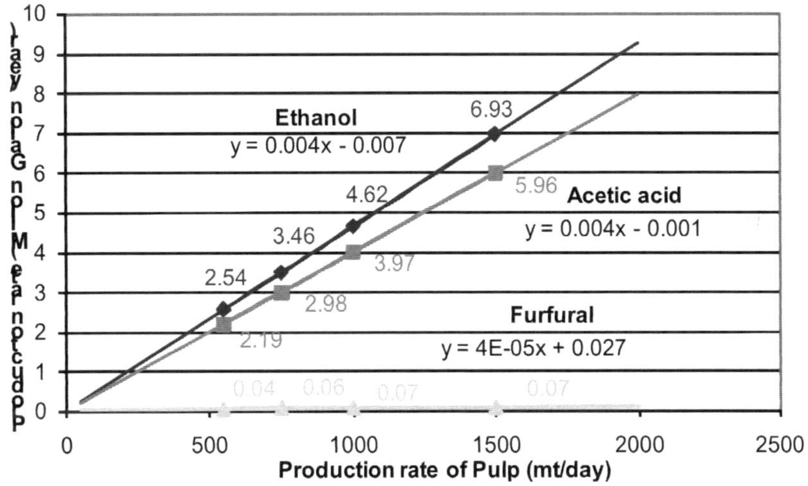

Figure 4 Products rate vs. plant size

Energy Consumption

The addition of the hemicellulose extraction process to a Kraft pulp mill reduces the energy which is obtained from the residual pulping liquor. No attempt was made in the present analysis to optimize the energy use by performing a pinch analysis [22]. The net energy output is illustrated in Figure 5 in terms of millions of BTU per hour as equivalent steam and electrical energy as a function of the pulp mill size. A comparison is made to the base Kraft mill case where no hemicelluloses are pre-extracted. The summary of the energy requirement for a 1,000 tonne per day pulp mill is illustrated in Tables 2 and 3. On average, the modified Kraft mill would produce approximately 35% less steam than the conventional Kraft mill because 10% of the wood mass is extracted and additional energy is required for the pre-evaporation and distillation operations [12]. Also, energy is lost during the extraction process because the extraction liquor is removed from the digester and flashed.

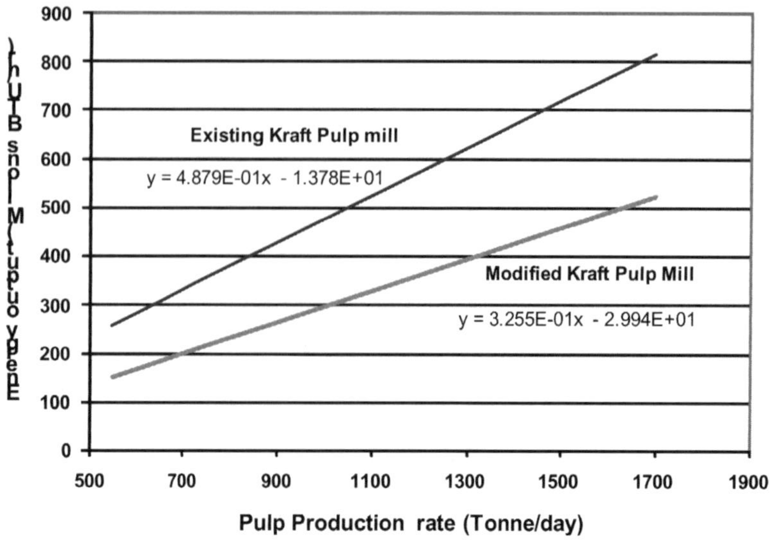

Figure 5 Energy output vs. plant size

Effect on Operation of the Lime Kiln

In the present situation the amount of Na_2CO_3 which needs to be processed in the lime cycle will be reduced by 20% for 30% sulfidity white liquor. This is equivalent to a reduction of 2.48% (as Na2O on dry wood). The oil consumption per tonne of dry lime mud (CaCO3) is 0.085 tonne of oil based upon the work of Gullichsen and Fogelholm [23], resulting in savings of about 9 tonne of fuel oil per day for a 1000 tonne per day pulp mill assuming a pulp yield of 47% and a causticizing efficiency of 80%. At a fuel oil price of about $2.2/gallon the oil saving are approximately $2 million dollars per year.

The savings in oil in the lime kiln of approximately $2 million dollars per year helps offset the loss of steam which is used to manufacture the new products. The value of the lost steam, assuming a steam cost of $7.00 per million BTU is $9.6 million dollars per year.

ECONOMIC ANALYSIS

The economic analysis was performed following standard methods described by Peters and Timmerhaus [24] and Turton et. al. [22]. In the analysis reported here it was assumed that an extraction vessel of some type was available at the site. For example in an existing Kraft pulp mill with two fiber lines in which one fiber line has been shut down, there may be an idle digester that could be converted into an extraction vessel and the plant operated as a single fiber line. A more likely situation in a single fiber-line pulp mill is that an impregnation vessel is available that could be converted into an extraction vessel since the wood must be impregnated with the pulping liquor in either case. In making the capital cost analysis, only 20% of the cost of a new continuous digester was applied to the project for modification of the impregnation vessel and/or connecting two continuous digesters in series. The capital cost estimates were obtained by scaling data provided by Wooley [7] and correcting for inflation and changes in unit price using the CEPCI index. In making the capital cost estimates a 15% contingency factor was applied to the estimated capital cost from the scaled data of Wooley [7]. Similarly, some mills would have sufficient waste treatment and utilities to accommodate the new process while other mills would have to expand their waste treatment and utility systems to account for the additional processing. Consequently two sub-cases were considered; one in which the utilities did not need to be upgraded and the case where they did.

Table 2
Energy Balance for Modified 1000 Tonne per Day Kraft Pulp Mill

Energy Balance Relative to Base Case				
Stream Number and Description		**Existing Kraft Pulp Mill Case**	**Modified Kraft Pulp Mill**	**Units**
Steam Required (Enthalpy)	**S3 Extraction**	0.0	84.6	**MMBTU/hr**
	S4 WL Preheater	20.9	15.6	**MMBTU/hr**
	S5 Upper Heater	84.6	41.5	**MMBTU/hr**
	S5 Lower Heater	41.8	64.7	**MMBTU/hr**
	S5 Washer Heater	29.6	41.1	**MMBTU/hr**
	S6 BL Evaporator	224.3	209.0	**MMBTU/hr**
	S9 Steam Dryer	68.0	68.2	**MMBTU/hr**
	S10 Extract Evaporator	0.0	0.0	**MMBTU/hr**
	S11 HAC Separation	0.0	11.8	**MMBTU/hr**
	S12 EtOH Distillation	0.0	12.7	**MMBTU/hr**
Total Steam Production (Enthalpy)	**Hogful Boiler**	129.4	129.1	**MMBTU/hr**
	Recovery Boiler	802.5	720.1	**MMBTU/hr**

Table 3
Summary of Energy Balance for Modified 1000 Tonne per Day Pulp Mill
Relative to Base Case

Stream Number and Description		**Existing Kraft Pulp Mill Case**	**Modified Kraft Pulp Mill**	**Units**
Steam Required in Extraction		0.00	84.6	**MMBTU/hr**
Steam Required in Pulping		469.2	440.1	**MMBTU/hr**
Steam Required in Ethanol Process		0.0	24.5	**MMBTU/hr**
Oil Savings in Lime Kiln		0.0	(15.3)	**MMBTU/hr**
Net Energy Output	**As steam**	462.7	300.0	**MMBTU/hr**
	As Electricity	27.1	17.6	**MW**

Note: Output of electricity was calculated by assuming 20% of the energy content in available biomass is converted to electrical energy.

Basis for Analysis

The capital cost was obtained from vender quotations on important equipment and by scaling cost data presented by Mitchell [8]. The basis for the economic analysis is summarized by Mao [12].

The plant was assumed to be constructed over a two year period. The selling price of ethanol was taken to be $2 dollars per gallon while the selling price of acetic acid was assumed to be $4 dollars per gallon. No additional wood was used in the process since experimental data showed that the overall pulp yield remains the same with the present "near-neutral" pre-extraction process. It was also assumed that the investment was completely borrowed and repaid over the life of the project which was taken to be 10 years. The capital recovery factor was assumed to be 15% based on a 10% interest rate; capital recovery factor of 16.3%. The viability of the project was determined by estimated the discounted cash flow rate of return and pay back period as described by Turton [22].

Capital Investment

In making the capital cost estimate, it was assumed in one version of the analysis that the waste treatment and utilities were insufficient at the mill site to support the new process and had to be upgraded. The relationship between size and Total Project Investment is shown in Equation 1 for the case where utilities and waste treatment had to be upgraded to account for the additional processing of hemicellulose extract. These additional charges accounted for approximately 20% of the capital investment. Total project cost for the three plant sizes considered is summarized in Table 4. It is clear from the equation that the capital cost increases as a logarithmic function with the size of pulp mill. In this power law the scaling exponent is equal to 0.684 for pulp mill production rates between 750 and 1,500 tonne/day. This scaling exponent is slightly larger than the normal 6-tenths rule. Total project costs were estimated to be between 30 and 47 million dollars depending upon the size of the pulp mill. The capital cost could be lowered if the utilities and waste treatment system were of sufficient size to handle the added processing.

$$\text{Total Project Cost (\$)} = 323{,}379 \times (\text{Plant size, Tonne/Day})^{0.684} \qquad (1)$$

Table 4
Estimate Capital Cost for the Near Neutral Hemicellulose Extraction Process

Pulp Mill Size (Tonne/Day)	Total Project Cost (Millions $)
750	29.5
1,000	36.2
1,500	47.3

Operating Cost

The operating cost was estimated as a function of the size of the pulp mill and is shown in Equations (2). The operation cost for the various plant sizes were estimated [Mao, 12] and fit to a linear equation.

$$\text{Operating Cost (\$)} = 13{,}512 \text{ (Plant Size, Tonne/D)} + 1{,}444{,}363 \qquad (2)$$

The unit production costs were estimated from the operating cost by assigning a portion of the operating cost to the production of both ethanol and acetic acid according to the amount of product produced on a mass basis. Using this simplified method of proportioning operating costs, approximately 43% of the operating cost was attributed to the production of ethanol and 57% to acetic acid. Table 5 illustrates the unit production cost for ethanol and acetic acid. Production costs for ethanol varied between $1.46 for the large pulp mill (1,500 tonne per day) and $1.56 per gallon for the small pulp mill (750 tonne per day). Similarly for the acetic acid the unit production cost varied between about $1.94 and $2.07 per gallon depending upon plant size.

Table 5
Ethanol Unit Production Cost

Pulp Mill Size (Tonne/Day)	Ethanol (Dollar/Gallon)	Acetic acid (Dollar/Gallon)
750	1.56	2.07
1000	1.51	2.01
1500	1.46	1.94

Discounted Cash Flow Analysis

An example of the discounted cash flow diagram is illustrated in Figure 6 for the 1000 tonne per day pulp mill for the case where the utilities and wastewater treatment system had to be upgraded. For this case the rate of return on investment was is 8.3%. Table 6 summarizes the discounted rate of return on investment for the different plant sizes. As the pulp mill size increases the discounted cash flow rate of return increase because additional product is produced and also because the operating cost per unit of product is lower.

Table 6
After Tax Discounted Cash Flow Rate of Return on Investment
Assuming Capital Investment Required for Upgrade of Waste Treatment System

Plant size (Pulp Production Rate)	Rate of Return on Investment
750 tonne/day	6.0 %
1000 tonne/day	8.3 %
1500 tonne/day	11.1 %

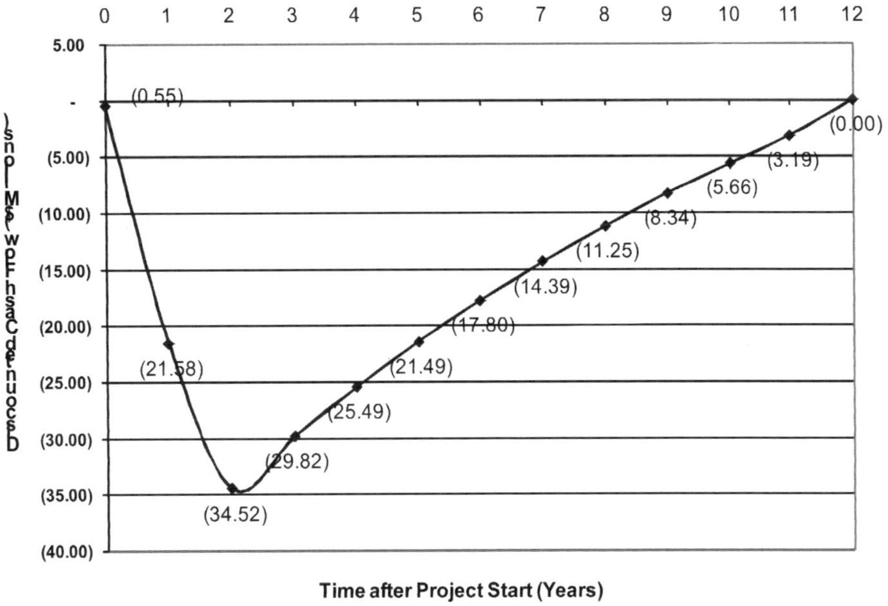

Figure 6. Discounted Cash Flow Diagram with Extraction Vessel Available for 1000 tonne per Day Case

Adequate Utilities and Waste Treatment

In the case where the utilities and waste treatment at the site were sufficient to handle the new process, the capital cost, unit production cost and discounted cash flow rate of return are summarized in Table 7. Approximately 20% less capital is required compared to the case where sufficient wastewater capacity and other utilities are available at the site; that is a total capital cost of approximately $23 million for the 750 tonne per day small pulp mill and $38 million for the 1,500 tonne per day large plant. For this case, the unit production costs are lowered to approximately $1.35 to $1.40 per gallon of ethanol and to $1.80 to $1.85 per gallon for acetic acid.

Table 7
Economy Analysis Basis and Results for No Waste Treatment System Upgrade

Plant Size (Tonne/day)	Capital Cost (Million Dollar)	Unit Production cost ($/Gallon)		DCFROR
		Ethanol	Acetic acid	
750	23.2	1.40	1.85	12.9%
1000	28.5	1.38	1.83	15.0%
1500	37.7	1.35	1.80	17.5%

CONCLUSIONS

The hardwood biorefinery considered here uses green liquor with the addition of AQ for wood extraction prior to modified kraft cooking to preserve both pulp yield and quality. During extraction about 10% of the wood goes into solution. The extract contains mostly hemicellulose derived organic compounds and has a near-neutral pH. The extracted chips require a 3% lower EA charge (as Na_2O on original oven dry wood) of white liquor and a lower H factor in the subsequent kraft cook as compared to a regular kraft cook.

In the present analysis, the extraction and pulping were assumed to be conducted using continuous processes. The assumption inherent in the analysis reported here is that an existing impregnation vessel can be converted into an extraction vessel. The rate of return on investment varies between 6% and 18% depending upon the plant size and whether the onsite utilities and wastewater water are of sufficient capacity to handle the additional processing from the hemicellulose extraction process. There are significant advantages to the new process, namely the pulp yield and quality are not degraded, the recovery boiler and the lime kiln are off loaded by about 20%, and two new by-products from the pulping process are generated. The off loading of the recovery cycle allows a potential increase in production rate if this section of the mill is forms the bottleneck for the production rate.

REFERENCES

[1] Wising, U., and Stuart, P., "Identifying the Canadian Forest Biorefinery", Pulp and Paper Canada 107(6), 25-30 (2006).

[2] Pervaiz, M. and Sain, M., "Biorefinery: Opportunities and Barriers for Petro-Chemical Industries", Pulp and Paper Canada 107(6), 31-33 (2006).

[3] Maybee, W. E. and Saddler, J. N., "The Potential of Bioconversion to Produce Fuels and Chemicals", Pulp and Paper Canada 107(6), 34-37 (2006).

[4] Magdzinski, L., Tembec Temiscaming Integrated Biorefinery", Pulp and Paper Canada 107(6), 44-45 (2006).

[5] Amidon, T. E. "The Biorefinery in New York: Woody Biomass into Commercial Ethanol", Pulp and Paper Canada 107(6), 47-50 (2006).

[6] Frederick, J.; Lien, S.J.; Courchene, C.; Ragauskas, A.J.; Iisa, K. Co-Production of Ethanol and Cellulose Fiber from Southern Pine: A technical and Economic Assessment. 2006 Engineering, Pulping, and Environmental Conference, Atlanta, GA. (Nov., 2006)

[7] Wooley, R., et al, "Lignocellulosic Biomass to Ethanol Process Design and Economics Utilizing Co-Current Dilute Acid Prehydrolysis and Enzymatic Hydrolysis Current and Futuristic Scenarios". NREL/TP-580-26157 (July 1999).
[8] Mitchell, J., "Production of Ethanol from Hardwood", M.S. Thesis, University of Maine (2006).
[9] Van Heiningen, A. "Converting a Kraft Pulp Mill into an Integrated Forest Biorefinery", Pulp and Paper Canada 107(6), 38-43 (2006).
[10] A. R. P. van Heiningen, H. Zou, S. Yoon, J. Jiang and G. Goyal, "Process of Treating a Lignocellulosic Material", U.S. patent application serial no. 11/640,820, December 18, 2006
[11] Sjostrom, E., "Wood Chemistry", p. 87, Academic Press, New York, 1981
[12] H. Mao. Technical Evaluation of Hardwood Biorefinery Using the Hemicellulose Extraction Process, Master's Thesis. University of Maine, 2007.
[13] WinGEMS help documents, Wingems 5.3 (Build 304) by Chemical Engineering Department at the University of Idaho.
[14] A. van Heiningen et al., University of Maine Forest Bioproducts R & D", Presentation given at 93rd PAPTAC Annual Meeting, Montreal, QC, Canada, February 7th, 2007
[15] Garrote, G. Dominguez, H. Parajo, J. C., "Generation of Xylose Solutions from Eucalyptus globules Wood by Autohydroysis-Posthydrolysis Processes Kinetics", Bioresource Technology 79(2001) 155-164.
[16] Aden, A., Ruth, M., Ibsen, K. "Lignocellulosic Biomass to Ethanol Process Design and Economics Utilizing Co-Current Dilute Acid Prehydrolysis and Enzymatic Hydrolysis for Corn Stover". NREL technical report, NREL/TP-510-32438, 2002. 27-36
[17] Amartey, S. and T. Jeffries. "An improvement in Pichia stipitis fermentation of acid-hydrolyses hemicellulose achieved by over liming (calcium hydroxide treatment) and strain adaptation", World Journal of Microbiology & Biotechnology Vol. 12, 281-283.
[18] Lawford, H. and Rousseau, J., "Fermentation of Biomass-Derived Glucuronic Acid by pet Expressing Recombinants of E. coli B", Applied Biochemistry and Biotechnology, 221-241 (1997)
[19] Molecular technology, http://www.ethanolindia.net/molecular_sieves.html
[20] http://www.ethanolrfa.org/industry/locations/
[21] http://www.the-innovation-group.com/ChemProfiles/Acetic%20Acid.htm
[22] Turton, R., et al., "Analysis, Synthesis, and Design of Chemical Processes", 2nd Ed. Prentice Hall, Upper Saddle River, N.J. (2003).
[23] Gullichsen, J and Fogelholm, C. J., "Chemical Pulping' in Pulp and Paper Making Science and Technology, Book 6B, pg B188. Fapet Oy Publisher (1999).
[24] Peters, M. S. and Timmerhaus, K. D., "Plant Design and Economics for Chemical Engineers", 4th Ed. McGrawHill New York (1991).

Process Modeling of Integrated Forest Biorefinery – A Comprehensive Approach

Hua-Jiang Huang, Weilu Lin, **Shri Ramaswamy**. Department of Bioproducts and Biosystems Engineering, University of Minnesota, 2004 Folwell Ave, Saint Paul, MN 55108

Recently there is an increasing interest in lignocellulosic biorefineries because of the rise in petroleum costs, the need for reducing greenhouse gas emissions, and the motivation to reduce our dependence on foreign oil. In the near future, integrated forest biorefineries (IFBR) based on conventional forest industries could become the most important lignocellulosic biorefineries due to its current capital infrastructure for converting renewable resources on a sustainable basis for over a century and help increase the renenue and products streams to improve their profitability. Here we present a comprehensive process model of the whole IFBR that will help analyze the overall performance including the economic and environmental factors.

The IFBR (Fig. 1) based on current infrastructure of pulp and paper mills involves pre-extraction of hemicelluloses prior to pulping (so called Value Prior to Pulping), isolation of short and long fiber after pulping, conversion of hemicelluloses and the short fiber (cellulose) to ethanol in one or more bioreactors, and the long cellulose fiber used for the production of paper and other fiber based materials such as bio-composites. In addition, the black liquor after pulping and washing containing dissolved lignin can be gasified to produce biosyngas substituting for the Tomlinson recovery boiler system. The resulting biosyngas can be further synthesized to produce chemicals such as methanol, liquid transportation fuels such as DME and green diesel, etc. Electricity and process steam can also be produced from biosyngas with high efficiency gas turbine.

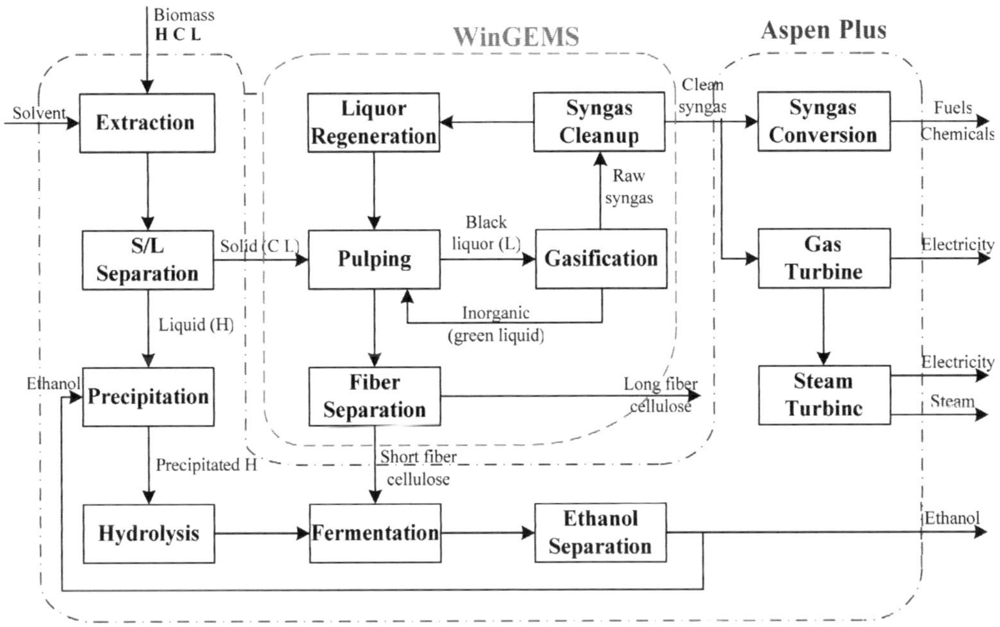

Fig. 1. Overall block diagram of a whole integrated forest biorefinery

The objective of the presentation is to develop a comprehensive approach for modeling IFBR, in order to achieve efficient design, simulation and optimization of the whole integrated forest biorefinery and to be able to perform the overall techno-economic and environmental analysis. An integrated method using two commercial process simulators Aspen Plus and WinGEMS and Microsoft Excel® as the communications interface between the simulators and a case study using the method are described. The comprehensive approach for modeling an integrated forest biorefinery and the mechanism of communications between Aspen Plus and WinGEMS with Excel as the communications interface is introduced with a case study.

Mechanism of communications between Aspen Plus and WinGEMS:

Briefly, data can be transferred between Aspen Plus simulation model and Excel by OLE technology, and between WinGEMS simulation model and Excel by the DDE technology. Thus, Aspen Plus can communicate with WinGEMS through Excel as a common interface/platform (Fig. 2). In other words, the simulation data from Aspen Plus model in an Excel spreadsheet and the simulation data from WinGEMS model in the same or different Excel spreadsheet can easily be shared with each other within Excel. For example, Aspen Plus is able to send a stream flowrate in the Aspen process model into a specified cell in an Excel worksheet from which WinGEMS retrieve the data to a specified stream in the WinGEMS model, and vice versa. Thus a seamless, efficient, real time data exchange between the two process simulators has been successfully achieved.

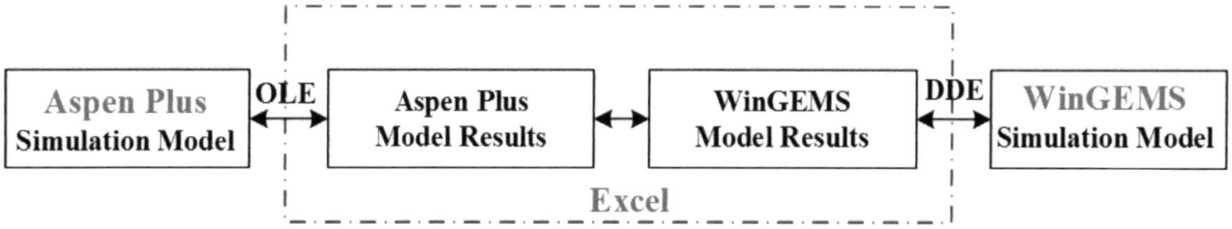

Fig. 2. Simplified diagram of communication between Aspen Plus and WinGEMS

The simulation results of a case study (details not shown in this extended abstract) show that the integrated method is efficient and reliable in modeling the simplified IFBR. The method presented here is equally applicable for modeling the complex IFBR processes including biomass-based integrated gasification combined-cycle (BIGCC) or black liquor gasification combined-cycle (BLGCC), biosyngas conversions into liquid fuels and chemicals, and fiber separation, etc. Efforts are currently underway in modeling the complex, whole integrated forest biorefinery with multiple products and co-products including energy and fuels. This general integrated approach will play an increasingly important role in the design and development of the future fully integrated lignocellulosic biorefinery of the 21st century bioeconomy.

DESIGN PHILOSOPHY OF ACTIVATED SLUDGE PROCESS IN EFFLUENT TREATMENT PLANT OF A KRAFT MILL BASED ON INDIAN MIXED HARDWOODS AND BAMBOO

Amiya K.Ray[1],Anirban Roy[2], Neha Bhaskar[2], Piyush Tiwari[2], mand Swodha Jha[2]
(1) Indian Institute of Technology Roorkee, Roorkee,India,
(2) Birla Institute of Technology,Noida Extension Centre,India

Abstract

An attempt has been made to design a mathematical model for an aerobic effluent Treatment Plant (ETP) using activated sludge process for process industry, like paper mill. Data from a running Indian paper mill using mixed hardwoods and bamboo are collected or measured, for simulation purpose. Various models and algorithms are developed for activated sludge process taking into consideration of bio-chemical kinetics and material balance relationship,for optimizing the dimensions of the aerator using optimization techniques such as Langrange multiplier,for estimating surface area of primary and secondary clarifiers of an ETP plant when the size of particle undergoing settling process is known and when SOR values are made available. Besides, an annual cost model is developed using the concept of Smith (3) for design of primary and secondary clarifiers and aerator considering all possible cost contributing factors for an entire ETP plant. The models are validated and simulation is carried out for general integrated paper plant with the data the data for a running paper mill in India.The developed software can be used for any process Industry if experimental and design data are made available.

Introduction

Activated sludge process belonging to slurry biological process group ,is used overwhelmingly in Indian process industries. Least cost optimization of a biological process with solids recycle is a complicated exercise because of the introduction of solids separation and recycle as variables. Basic process performance in terms of specific effluent quality can be maintained across a wide spectrum of combinations of reactor volume, clarifier area, and recycle policy because the biological solids retention time is no longer equal to the hydraulic retention time. Smith(1) has reported data on conventional and advanced treatment waste water.Steady state material balance on activated sludge process was put forward by Ramalho (2) related to only BOD removal. Critical reviews of optimal design have been depicted in various design handbooks(2-5). No work is attempted to design based on removal of other pollutants like COD, AOX or TOCl. Many other reports are also available(6-7) for design and control of activated sludge process for pulp and paper and other plants. In practice, it is extremely difficult to precisely find out the total treatment cost including fixed charges due to capital investment for equipments for a Greenfield mill or a mill under operation as accurate value can only be obtained if detailed accounts of pulp types, raw materials, washing practices, bleaching sequence, and specially the ETP configurations (pre, primary, secondary and tertiary treatment, aerobic/ anaerobic or their combination) are known. Utilizing the above approaches, it is possible to develop systematic design procedure for biological process and the system as a whole and an order of estimate values can always be worked out for any industry within some broad range of data. In the present investigation , an attempt has been made to design a activated sludge process of waste water treatment from an Indian paper mill based on wood and bamboo at a specified level of effluent quality.

Development Of Mathematical Models and Algorithms

Activated sludge process is an important equipment in effluent treatment plant of process industry including pulp and paper. For a computerized design of an ETP plant , various models and corresponding algorithms are developed as follows:

Algorithm-1 Steady state material balance relationships with kinetic rate and flocculation in aerator (2).

The algorithm for detailed calculation based on BOD has already been developed (2) which consists of several steps such as calculaion of the recycle ratio of MLVSS based on Effluent Quality Controls Design or based on Optimum Flocculation Conditions Design. This will follow for the estimation of BOD of the combined feed,$S_o = (S_F + r\ S_e) / (1+ r)$. The other parameters of the sequential design are: residence time,F/M(0.6>F/M>0.3), net yield of MLVSS, oxygen requirements,neutralization and nutrient demands,sludge age,volumetric sludge load and circulation number. These are defined as follows:

Sludge age, the mean residence time of MLVSS in the reactor: T_s = lb MLVSS in reactor/net output of VSS from the system

Volumetric sludge load,$S_S = S_H*C*SVI$

Checking circulation Number, CN(within the range 80-90%) is also important. This is related to Q_F and Q_w as follows:

$Q_r = rQ_F)$;$Q_w = Q^{//}$;$CN = (Qr/Q_w) +1$

Cross- checking the component material balance for MLVSS and the overall material balance for NVSS are essentially required.

For the design of the system the following must be adopted:

Material Balance Calculations:

Throughput Rate= $Q=Q_F+Q_R=Q_F\ (1+r)$

Equipment specifications:

The reactor volume, reactor depth and cross sectional area

$V=Q^*\ \tau= Q_F\ (1+r)\ \tau$

Lagrange multiplier for optimization of dimensions of aerator (**see algorithm-2**) has been used in the present study.The depth has been optimized which normally lie in the range: 10-15 ft (3.048-4.572 m).

Specification of aeration equipment:

Development of models for optimization of ETP equipments:

The development of a complete model requires modeling of aerator, primary and secondary clarifier plus other auxiliary equipments. Separate algorithms thus have to be developed.

Algorthm-2 Optimization of aerator (reactor) dimensions by Langrange multiplier and Newton-Raphson method

The detail of the mathematical theory of Lagrange multiplier and Newton –Raphson method is given in any standard texts of mathematics and operation Research or optimization techniques. The tank volume = lbh, where l is the length, b is the width, and h is the depth to minimize the surface area, i.e. the function

F (l,b,h) =2 hb+lb+rhl
Constraints: f (l,b,h,) = 0, i.e f(l,b,h) = lbh-V
The auxiliary function has the form
$\varphi = (l, b, h, \lambda) = 2bh+lb+2hl+\lambda\,(lbh-V)$
$\varphi = (l, b, h, \lambda) = 2bh+lb+2hl+\lambda\,(lbh-V)$
On taking partial derivatives with respect to l, b, h, and λ
$\partial\varphi/\partial l = b + 2h + \lambda\, bh = 0$ .... (a)
$\partial\varphi/\partial l = 2h + \lambda 1h = 0$ .... (b)
$\partial\varphi/\partial h = 2b + 2l + \lambda lb = 0$ .... (c)
$\partial\varphi/\partial h = lbh-V = 0$ .... (d)
Setting these four partial derivatives equal to zero and solving the four equations simultaneously, from Eq. (a) one gets
$\lambda = -(b+2h)/bh$ .... (e)
Substituting the expression fro λ in Eq. (b) and solving it for b, gives b = 1 by Newton-Raphson method.
From Eq. (c) one can find that $\lambda = -4/b$, and from Eq. (a), it follows that b = 2h

Thus $V = lbh = (b)\,(b)\,(b/2) = b^3/2 = m^3$ when 1= m,b = m, and h = m
To find the area of cross section of a clarifier from the data of mass flow rate, kg/s, initial and final concentration of suspensions (by mass), density of clarified liquid and actual settling velocity, one can use Algorithm-3 and Algorithm-4 as follows: For complete cost estimation cost of primary and secondary clarifiers along with the costs of piping, pumps and motor are to be determined. These are shown in the algorithm-4.

Algorithm- 3: To develop a model for finding the area of primary and secondary clarifier and their dimensions (diameter and height) if minimum size of suspended particle(d_{min}) in clarifier is known.

This algorithm also helps to estimate the height of a clarifier. This algorithm needs to determine the various flows in and out of a clarifier together their concentration terms. The algorithm is however valid for a circular geometric design. Use Algorithm-3 for both primary clarifier and secondary clarifier using respective values of all parameters.
The material balance relationship developed is as follows:

Material Balance for the Clarifiers (primary/secondary):

Using Q,the flow,C,the concentration, ρ,the density and the subscripts,f,u and F are for feed,underflow and overflow respectively, one can write the following.

Overall balance:

$Q_f \rho_f = Q_u\rho_u + Q_F\rho_F$

Solid balance:

$Q_f\, C_f = Q_uC_u + Q_FC_F$

For ideal case,

$Q_F\, C_F = 0$ as C_F is assumed to be zero

Water balance:

$Q_f \rho_f - Q_f\, C_f = (Q_u\rho_u - Q_uC_u) + (Q_F\, \rho_F - Q_FC_F)$
$Q_f\, (\rho_f - C_f) = (Q_u\rho_u - Q_uC_u) + (Q_F\, \rho_F - Q_FC_F)$
$Q_f = = [(Q_u\rho_u - Q_uC_u) + (Q_F\, \rho_F - Q_FC_F)]/\,(\rho_f - C_f)$

For ideal condition:

$Q_f = = [(Q_u\rho_u - Q_uC_u) + Q_F\, \rho_F)]/ (\rho_f - C_f)$

$Q_f = [Q_u\, (\rho_u - C_u) + Q_F\, \rho_F]/ (\rho_f - C_f)$

To determine the value of Q_f, one has to know the values of Q_u, Q_F, ρ_f, Cf, ρ_u, Cu

From solid balance equation, one can get fairly accurate values of Q_u, and Q_f from the data of Q_f, C_f and C_u as follows:

Assuming there is no solids in the overflow ($C_F \approx 0.0$). Though in ideal design it is true, in actual practice, this is not so. However, the impacts of small concentration of overflow solids on the flow streams are not significant and this is allowed for preliminary estimates to start with in engineering calculations.

$Q_f\, C_f = Q_uC_u$

$Q_u = Q_f\, C_f / C_u$

Now this equation can be used for dimensioning of both primary and secondary clarifier by changing the values of the parameters.

The general algorithm can be developed which consists of calculating settling velocity of the particles,checking the Reynold's Number,calculating actual settling velocity, finding out the area of primary clarifier A_p, and that of secondary clarifier, A_S,determining the diameter of the clarifier,D,relative density of suspension,Δ_{susp},calculating mass concentration of the suspension in the compression zone, x_{sc},estimating the amount of solid phase per m^3 of suspensions, C_{sc} and the amount of solid phase settling on per day per1 m^2 basis. Knowing SPPDSM based on initial conditions, and the amount of solid phase per m^2 of settling area (SPPSM), the height of the compression zone, h_2 can be estimated.Now adding the height of the feed zone, h_1, the height of the settling tank zone in which the blade rotates depending upon the inclination of the blade relative to the tank bottom, h_3 ,the total height, h_t can be estimated.

Algorithm-4: To develop a model for finding the area of primary and secondary clarifier if minimum size of suspended particle ($d_{min.}$) in clarifier is not known but surface overflow rate (SOR) is made available

Algorithm-3 is useful when the particle size undergoing sedimentation/settling is known and Stokes law is applicable. This is feasible when, there is no layer sedimentation, concentration of particles of known density is less, individual particle having uniform size and shape, settles from rest without hindrance of neighboring particles,particle density is known

Though the model is suitable for finding out the height of clarifier (thickener) it can not be applied to the calculation of clarifier in ETP mostly because of want of particle size and its distribution. Practical designers take help of the design value of surface overflow rate (SOR) for preliminary design estimates.

SOR is varying between 800-1000 gallons/sq. ft day for primary clarifier and 600-800 gallons/sq. ft day for secondary clarifier

The procedure of step by step calculation can be shown in the algorithm which has further been extended to examine the relative and annual cost of an entire plant. For complete economic description of an ETP, an overall annual cost model is to be defined. The various information regarding cost indices are taken from various design and data handbook (3,4,5).

Algorithm-4:Development Of Objective Function, C_t By Modeling Annual Cost Equation Of A General ETP Plant

Based on the cost equation developed by Smith (1,6) and further reproduced by Mishra et al. **(7),** the economic models in the present investigation are developed. Some simulation work is also presented by Mishra et al. (6) to examine the effects of flocculation in primary clarification unit and the overall impact of putting a primary clarifier on capital investment. For cost data upgradation, the guidelines of Perry(5) and Peter and Timmerhaus(3) are followed.

A. For a Greenfield mill:
$(C_T)_{min}$ = Capitalized cost for primary clarifier (C_{pc}) + Aerator (C_a) + secondary clarifier (C_{sc}) + other engineering expenses

B. For complete design of ETP of an already existing mill :
$(C_T)_{min}$ = Fixed charges due to primary, and secondary clarifier and aerator, C_{F1} + annual power cost duc to aerators, C_{powa} + power cost due to rake movement in primary and secondary clarifier, C_{pps} + total annual pumping costs, C_{pump} + annual cost of all pipes $C_{pipe\ 3}$ + Fixed charges for pump and motor, C_{PEM2} + cost of nutrients, C_{nut} + operational charges due to labour, C_L.

The above models need to calculate the area of primary,A_{pl} and secondary clarifierA_{s1}, volume of the aerator, V_A, MG (from ALGORITHM-1), their proportional cost using Williams' Sixth tenth factor, upgradation of the cost with price index ratio for present worth. These are as follows:

Q_F = TPD* q ; Q_F* C_f = Q_u*C_u;Q_F = Q_f-Q_u; Q = Q_F (1+r)

A_{p1}, = TPD* q/SOR;A_{S1}= Q*10^6/SOR

Taking r from first algorithm (r≈0.36) one can get,

Q = Q_F (1+r), Q and Q_F in MGD

V_{A1} = Q*t = Q_F (1+r)*t; (t in day^{-1}); Q in MGD

Cost of aerator, C_A, US$: $C_A = V_A\,[175{,}000 + 36{,}500\ V_A^{-0.818}]$, V_A, MG

In the present study ,C_f = 0.552 kg/m^3, C_u= 15.2kg/m^3, r can be varied between 0.0-1.0.

The model can be subjected to various situations and can be solved easily:

Total investment at various values of recycle ratio, r can be estimated by iterative procedure.

The volume of aerator can be found out by by optimization techniques by minimizing the total annual cost of ETP plant.

The area of clarifiers can be estimated through optimization techniques by minimizing the total annual cost of ETP plant.

Algorithm-5:Modeling For Finding Optimum Economic Diameters Of All Pipe Lines Of ETP

The investment for piping and pipe fittings of an ETP plant can amount to an important part of total investment for a chemical plant which demands optimum economic pipe diameter calculations. The optimum economic pipe diameter can be found at the point at which the sum of pumping and fixed charges based on the cost of the piping system is a minimum. Economic pipe diameter calculation has been made based on the information available in existing literature(9,14) with the consideration of flow regimes(laminar or turbulent). The entire algorithm has been solved based on iterative procedures with the following check values

- Compare the velocity values with the design value
- It must match within 10% of the given value

Model Validation And Simulation

Model-1-5 are validated using known data of the waste water treatment plant from municipal wastes(2) and then extended to pulp and paper plant for simulation(4). Model-5 is simulated using the data of Peter and Timmerhaus(3) and Perry(5) for cost estimates of optimal economic pipe lines and pumping costs. Cost indices are considered based on the average value of Marshall & Swift and Chemical Engineering plant cost index(3).Average standard cost of thickener,costs of pump, and motor etc. are taken from Perry(5). The parameters influencing the capacity and efficiency, such as content, ash content, and pH, MLVSS, BOD_3 and COD from primary,secondary and aerator and sedimentation rate from primary and secondary clarifier (inlet, and outlet) are measured experimentally in the laboratory.Some other variables like F/M ratio, reycle ratio, residence time, nutrients amount are obtained from the industry. Design parameters such as pipe diameter and length of various pipelines, power of aerators, pump capacity and their power of pump and their specifications such as capacity, flow rate etc. are measured in the industry. Detailed calculation has been attempted for various plant capacities having nearly 150-200 TPD capacity. For conversion of US$ to Indian rupees (INR) are taken from published data. Now the models can be used to examine the effect of plant capacities on the economic design of ETP.

Constraints of ETP in Indian mill

Basis:10 MGD,TPD=200,q=200 m^3/t)**:** F/M = 0.2 less than the range:0.6-0.7 < F/M >0.3 Env.Parameters(mg/l):5% of $S_F<S_e>10\%$ of S_F ;2000 $<X_{v,a}>3000$;SVI<300; 10,000 <Xv >15,000 ; r = 0.1946-0.275; S_e = 20-30 (avg. 24.8); $X_{V,a}$= 1500;$X_{NV,a} = X_{NV,o}$= 300;$X_{NV,u}$ = 10000;$X_{V,e}$ = 0;S_F = 370

System	PH	Retention Time,h	TDS,ppm	S.S.,ppm/mg/l	COD, mg/l	BOD, mg/l
Primary	Input: 8.2 Outlet:7.4	10	Input:1626 Outlet: 1556	Input:552(**$X_{NV,F}$**) Underflow:15200 Outlet :75	1062	370
Secondary	Overflow:7.6	6	Outlet: 1480	Underflow: 5025 Overflow: 81.2	586	230-300
Aerator	7.5-7.7	15	-	MLSS=1948-1582(Avg. 1813) MLVSS = 1544-1282(Avg. 1412)	203	24.5-28

Results And Discussions

The developed models and the algorithms (Algorithms 1-5) are now subjected to a set of computer programs on C++ language. These programs are as follows: Algorithm-1 finds out the recirculation ratio, r, residence time, F/M ratio, reactor volume, outlet steams from secondary clarifier, Oxygen requirement, power consumption, neutralization requirements, sludge age, volumetric sludge load and circulation number for any industry including pulp and paper waste water treatment plant.The results are shown in **the Table 4.1** for Q_F varying between 2.5 MGDand 7.5 MGD.

The effects of changing the capacity from 1.5 to 10 MGD on volume, and dimensions of aerator are shown in the **Table 4.2.**

Algorithm - 2 finds the optimum dimensions of aerator estimated from the programs. **Table 4. 3** depicts the computed results

Algorithm - 3 gives a general methodology for finding out the area of clarifiers and the height of compression zone and also the total height if the size of particle undergoing settling is known. This can be used for any industry and for any process when discrete particle sizes are accurately measured by experiments or estimated.

Algorithm - 4 is developed which can be used for a number of cases

The model is examined if one desires to see the impact of changing capacity on capital investment and operational cost. The results are shown in Table 4. 3.

Table 4.1 Estimation of various parameters from material balance of various plant capacities

Parameters	**2.5 MGD**	**5 MGD**	**7.5 MGD**
Recirculation, r	0.359018	0.359018	0.359018
Residence time, t	0.589067	0.589067	0.589067
Reactor Vol., V	267564.19	535128.39	802692.63
F/M Ratio	0.505635	0.505635	0.505635
Q''	0.121719	0.243438	0.365157
Q'	2.378281	4.756562	7.134843
ΔX_{NV}	228.8027	457.6054	686.4081
ΔX_t	16380.16	20760.32	31140.49
Oxygen Reqd.	266168.12	53236.23	79854.36
Power Reqd. HP	499.08425	998.1685	1497.253
No. of Aerators	09	19	29
Neutralization	NO	NO	NO
Sludge Age	4.932791	4.932791	4.932791
S_{SVI}	14.6664	14.6664	14.6664
Circulation No.	8.373409	8.373409	8.373409

Table 4.2 Results of optimizing the aerator as a function of capacity:

Capacity, MGD	**Volume, V_A ft^3**	**Length, a, ft**	**Breadth, b, ft**	**Height, c, ft**
2.5	267564.198	81.18689	81.18689	40.59344
5.0	535128.397	102.2890	102.2890	51.14453
7.5	802692.633	117.0917	117.0917	58.54587

Table 4.3 Comparison of capital investment for three major equipment in ETP plant of a paper mill

Plant Cap., TPD	**Pri. Cl., C_{p11}, %**	**Sec. Cl., C_{S11}, %**	**Aerator, C_{a11}, %**	**Total**
200	47.21	9.91	42.88	100
300	42.22	8.87	48.91	100
400	38.66	8.12	53.22	100

It is indicated that with the increase of plant capacity the costs of both primary and secondary clarifiers in terms of percentage decrease. In sharp contrast the same for the aerator increases slowly.It is also evident that while for low capacity the costs of primary clarifier and aerator are nearly the same, but the cost increases for aerator as the plant capacity increases in much greater proportion.It is also reflected that the cost of secondary clarifier are much less (less than 10%)in comparison to the costs of primary clarifier and the aerator.

Table 4.4 compares the cost of various equipments , pipe and pump of an ETP plant of a paper mill using hardwoods and bamboo as a function of plant capacity.

Table 4.4 Comparison of capital investment of all equipments:

Plant Cap., TPD	Cost of Pri-Sec-Aer., $C_{psae\ I}$, %	Cost of pipe, $C_{pipe\ 1}$, %	Cost of pump, $C_{pem\ I}$, %	Total
200	95.950	1.324	2.726	100
300	96.003	1.558	2.439	100
400	96.033	1.733	2.234	100

It is evident from the table that the major cost for capital investment is due to the construction of primary and secondary clarifier and aerator. This amounts to approximately 96% ,relatively constant, and independent of capacity. The cost of pump and pipe lines are very small. However the later is found to be almost half of the former. With the increase of capacity the combined cost of primary and secondary clarifier and aerator decreases, cost of pipe line increases while the cost of pump decreases.

Table 4.5 compares the various cost components of the same plant as a function of plant capacity.

Table 4.5 Comparison of contribution of operational charges:

Cap., TPD	$C_{f\ I}$, %	$C_{pow\ I}$, %	$C_{pps\ I}$, %	$C_{pump\ I}$, %	$C_{pipe4\ I}$, %	$C_{pem2\ I}$, %	$C_{nut\ I}$, %	$C_{L\ I}$, %
200	8.620	85.792	0.054	4.960	0.285	0.245	0.027	0.015
300	6.674	89.105	0.037	3.725	0.260	0.169	0.019	0.100
400	5.581	90.975	0.029	3.021	0.242	0.129	0.015	0.008

The **Table-4.5** indicates that the cost due to power consumption is very large which has the largest share compared to other cost contributing factors. The next higher value is due to fixed charges and repair – maintenance cost of primary – secondary clarifier and aerator. The cost due to pumping follows the next. The cost of, power consumption due to rake movement, labour and nutrients are insignificantly small compared to others. Fixed charges due to pump and pipe line are of the same order. The power cost increases with the increase of plant capacity whereas the fixed charges decreases in all the cases.The total capital cost and operational charges as a function of plant capacity is shown in the Table 4.6 which shows that the annual operational charges is much more than that of capital investment.As expected, all the cost terms increases with the increase of plant capacity. The plant capacity of an existing paper mill using hardwood and bamoo is on the order of 150-200 TPD.The design of an existing plant of 10MGD is simulated. The results of simulation is shown in Table 4.7

Table 4.6 Capital cost and operational charges for various plant capacities

Plant Cap., TPD	$CT_{capital}$, Rs.	CT_{psa}, Rs.	CT (oper.), Rs
200	$5.8968*10^8$	$4.041*10^8$	$1.3127*10^9$
300	$6.5896*10^8$	$4.519*10^8$	$1.8958*10^9$
400	$7.1945*10^8$	$4.935*10^8$	$2.476*10^9$

Table 4.7 Comparison of Design specification of an existing pulp and paper mill

Items	A_p,ft^2	A_s,ft^2	V_A, MG	D_i, inch	D_{up}, inch	D_{us}, inch	Cap. MGD
Design	20108.55	17672.08	6.25	24	6.0	18.0	10.0
Present	13208.6	17311.4	6.294	23.46	5.78	17.54	10.18
Error%	-34.31	-2.04	+10.784	-2.25	-3.67	-2.55	+1.8

The present model predicts the design remarkably very well for all the data of volume,area, diameters of clarifiers and aerator except the primary clarifier. The present calculation predicts the capacity, area of the secondary clarifier, diameter of the of inlet pipe, diameter of the outlet pipe from the primary clarifier, and diameter of outlet of secondary clarifier very accurately with % error 1.18%, -2.04%, -2.25%, -3.67%, -2.55% respectively. Using 525 gallons/ft^2 day the value of SOR for primary clarifier, area becomes 20126.5 ft^2 and the corresponding % error will be of the order of 0.089%.

Conclusion

In this present investigation, a procedure for design of an ETP for any process industry using activated sludge process has been developed.Though the design and simulation is based on the effluents from a paper mill based on hardwood and bamboo,it can also be applied to municipal waste and other process industries like sugar, textile, and distillery. For this purpose various models and the corresponding algorithms are developed.A model for optimizing economic pipe diameter is taken from Perry (5) and Peters and Timmerhaus (3) and modified accordingly for simulation.The simulated data based on models predicts reasonably well for all the equipments and accessories of ETP plant. The constraints for primary clarifier in Indian industry justify its value predicted from the model.

References

1. Smith, R "Cost of Conventional and Advanced Treatment Waste Water," J. Water Poll. Control Fed. 40 1546 (1968).
2. Ramalho, R.H.- Introduction to water treatment processes-Academic press, 1976
3. Peters, M.S and Klaus D. Timmerhaus "Plant Design and Economics for Chemical Engineers- International Students Ed., McGraw- Hill Kogakusha Ltd., London, 1980
4. Springer Allan M.- Industrial Environmental Control- Second Ed.TAPPI,Atlanta, 1993
5. Perry R.H.,and Cecil H. Chilton-Chemical Engineers' Handbook-5thand7th Ed.,McGraw Hill. 1973
6. Smith, R.- JWPCF, 40(9), 154(1968)
7. Mishra N.P., Liang – Tseng Fan, L.E. Erickson and Ming-Ching T.Kuo-Simulation Studies of the Activated Sludge System including a Primary Clarifier-II Ch, E, 1973Vol.XV(3), pp-18

533b Solid And Microcellular Polylactide-Carbon Nanotube Nanocomposites

Shaoqin (Sarah) Gong[1], Srikanth Pilla[1], Adam Kramschuster[2], Alexander Chandra[2], and Lih-Sheng Turng[2]. (1) Mechanical Engineering, University of Wisconsin-Milwaukee, 3200 North Cramer Street, Milwaukee, WI 53211, (2) Department of Mechanical Engineering, University of Wisconsin-Madison, 1513 University Avenue, Madison, WI 53706-1572

Poly(lactide) (PLA)-multiwalled carbon nanotube (MWCNT) nanocomposites were melt-compounded using a twin-screw extruder. Solid and microcellular tensile bar specimens were produced via conventional and microcellular injection molding processes, respectively. Various characterization techniques were applied to study the static and dynamic mechanical properties, degree of MWCNT dispersion, cell morphology, and crystallization behavior. The addition of a small amount of MWCNTs led to a decrease in the cell size and an increase in the cell density of the microcellular PLA. A transmission electron microscopy analysis of the PLA-MWCNT specimens revealed a higher degree of MWCNT dispersion in the microcellular PLA-MWCNT composite compared with its solid counterpart indicating the microcellular injection molding process further dispersed the MWCNTs. While the addition of 1.5 wt% MWCNTs slightly increased the average specific modulus of the microcellular PLA, it did not affect the average specific modulus of the solid PLA. In addition, for both solid and microcellular specimens, the specific strength, specific toughness and strain-at-break decreased with the addition of MWCNTs. The storage modulus was not affected significantly with the addition of MWCNTs, but was found to be higher for the microcellular specimens compared with their solid counterparts. Finally, the crystallinity of PLA increased significantly with the addition of MWCNTs.

533c Simultaneous Heat And Mass Transport In Paper

Bandaru V. Ramarao and Sergiy Lavrykov. Empire State Paper Research Institute, State University of New York, 1 Forestry Drive, Syracuse, NY 13210-2278

Simultaneous heat and mass transfer through paper is considered.

533d Enhanced Bondability Of Inorganic Particles With A Cellulosic Substrate By Encapsulation With Regenerated Cellulose

Yulin Deng and Kimberly Nelson. School of chemical and bio-molecular Engineering, Georgia Institute of Technology, 500, 10th st. N.W., Atlanta, GA 30332

New insights into the solution structure and precipitation behavior of cellulose dissolved in 4-Methylmorpholine N-oxide (NMMO) monohydrate solvent were obtained from rapid precipitation of the cellulose aggregates upon addition of a non-solvent under agitation. Regenerated cellulose nanoparticles were synthesized with controllable morphology, crystalline phase, and crystallinity. Depending on the rate of non-solvent addition, the regenerated cellulose formed wormlike microfibrillar ribbons or spherical nanoparticles with diameters from 20 to 150 nm. Inorganic-cellulose core-shell and hollow cellulose nanoparticles were also prepared by self-encapsulation with regenerated cellulose via precipitation of cellulose in a polyacrylic acid hydrogel layer surrounding nanoparticle templates in NMMO. The cellulose shell surrounding the inorganic surface is composed of ~30 nm cellulose spheres or nodules. It was found that the bondability of the inorganic filler with wood fibers could be significantly improved by coating regenerated cellulose on it, and the degree of this improvement is dependant on the crystallinity, and therefore swellability and conformability of the coating layer.

Exploring the potential of Lantana Plant for pulp production in Indian Mills-An Experimental and Statistical Approach

Amiya.K. Ray and Manish Kumar Puri,, Indian Intitute of Technology Roorkee,India

Abstract

Indian paper industry is passing through a crisis due to shortage of raw materials, high cost of energy, huge consumption of water and power and generation of large amount of pollutants. Among these all, raw material shortage is one of the major problems. Paper industries are always in search for possible non wood fibrous raw material that can replace traditional wood and bamboo. Lantana camara, a plant of *Verbenaceae/ Vervain* family, having 75.03% hollocellulose, 8.461% extractives, 18.21% lignin and 2.31% silica can be a good raw material for paper making. In this work the potential of Lantana camara is evaluated for papermaking. The raw material is however renewable. Since the raw material is new and no correlations are available to evaluate the results, here in this paper a statistical model has been developed to predict the yield, kappa number, brightness and lignin content through statistical experimental design.

1. Introduction

In India Lantana is abundantly available, not only limited in Himalayas region but also in plain land. It is widely distributed around Saharanpur, Hardwar, Najimbabad, and Bijnore. The Lantana normally available in India is of the type of species Lantana camara or Lantana scandes. The plants origin is West Indies, and belongs to the family *Verbenaceae/ Vervain.*

Lantana camara is a perennial shrub. Some plants grow to as much as 5 to 6 feet tall, forming large, bushy mounds while others stay low and spreading, reaching up to 4 feet wide, but only 1 to 2 feet in height. Flower colors vary from pale cream to yellow, white, pink, orange, red, lilac and purple. Original forms have glossy rounded fruits that are fleshy and purplish-black when ripe.

The constitutes of leaves of Lantana camara is being used as medicinal source. These are Eight triterpenoids, betulonic acid ,betulinic acid ,oleanolic acid , lantadene A ,lantadene B, icterogenin, lantanilic acid , and ursolic acid , three flavonoids , hispidulin , pectolinarigenin , and pectolinarin , as well as β-sitosteryl-3-*O*-β-Dglucoside and a mixture of campesterol , stigmasterol , and β-sitosterol were isolated from the leaves of the yellow flowering *Lantana camara*.The structures of these compounds were established by spectroscopic methods.

The material can be exploited commercially for production of pulp in one hand and recover the components from leave and bark of the plant for medicinal use in other. In this present investigation an attempt has been made to produce pulp using Kraft pulping from the wild shrub locally available at Saharanpur. Further to do the exploiting examination of Lantana a number of variables required to be optimized. The variables are, time at temperature, sulphidity and time. Bath ratio and sulphidity was kept constant at 4:1 and 17% respectively. The statistical experimental method followed in the investigation is described below.

2. Statistical Design Of Experiments

Many pulping experiment are of extremum problem type e.g. determination of optimum conditions of a process, finding of optimum constituents of a compound, etc. In such cases to effectively tackle the problem in an economical way, a systematic design of an experiment is selected. The investigation scheme is usually broken down into some individual steps to evaluate the strategy for further experimentation by using experimental results obtained from each previous step. Thus the experimenter is able to conduct the experiment in an optimum way.

With design of experiment, it is possible to vary all the factors simultaneously and obtain quantitative estimate of the main effects and factor interactions among all the input and output variables can also be established.

2.1 Full factorial design

In this investigation statistical model is to be developed which requires to be verified by empirical data to predict the use of process design to set unified model using full factorial design technique. Keeping the above in view following stepwise investigation has been carried out.

The number of factors chosen was three with three numbers of levels. The total number of experiments thus to be carried out 27.The factors were symbolized temperature (Z_1), time (Z_2) and sulphidity (Z_3).

The above Z j variables are transformed into a dimensionless system of co-ordinates Xj.

The lower level Z_1^{min} and the upper level $Z_1^{\ max}$ will give the values of coded factors:

$$Z_1^{\ 0} = \frac{Z_1^{\ max} + Z_1^{\ max}}{2} \qquad (2.1) \qquad\qquad \Delta Z_1 = \frac{Z_1^{\ max} - Z_1^{\ min}}{2} \qquad (2.2)$$

The points with coordinates (Z_1^0, Z_2^0 ,.......$Z_k^{\ O}$) are called centre point of the design or the basic level, ΔZj is the unit or interval of variation on the Zj axis. It is usual to pass from the Z_1,Z_2,Z_3,Z_4.........Z_k coordinates to a new dimensionless system of coordinates x_1,x_2,x_3.......x_k by the coding equation

$$x_j = \frac{z_j - z_j^{\ 0}}{\Delta z_j} \qquad (2.3)$$

In the dimensionless coordinates system, the upper and lower levels are at +1 and -1.X_0 is dummy variable taken as 1.

Table 3: Design matrix with coded factors:

Time	Temperature	Sulphidity	Xo	X1	X2	X3
1.5	150	15	1	-1	-1	-1
1.5	150	17	1	-1	-1	0
1.5	150	19	1	-1	-1	1
1.5	160	15	1	-1	0	-1
1.5	160	17	1	-1	0	0
1.5	160	19	1	-1	0	1
1.5	170	15	1	-1	1	-1
1.5	170	17	1	-1	1	0
1.5	170	19	1	-1	1	1
2	150	15	1	0	-1	-1
2	150	17	1	0	-1	0
2	150	19	1	0	-1	1
2	160	15	1	0	0	-1
2	160	17	1	0	0	0
2	160	19	1	0	0	1
2	170	15	1	0	1	-1
2	170	17	1	0	1	0
2	170	19	1	0	1	1
2.5	150	15	1	1	-1	-1
2.5	150	17	1	1	-1	0
2.5	150	19	1	1	-1	1
2.5	160	15	1	1	0	-1
2.5	160	17	1	1	0	0
2.5	160	19	1	1	0	1
2.5	170	15	1	1	1	-1
2.5	170	17	1	1	1	0
2.5	170	19	1	1	1	1

Step3: A design matrix is transformed to experimental matrix which is the matrix formed out of coded factors.

Step 4: A suitable regression equation is assumed of the form

$$Y=b_0+b_1X_1+b_2X_2+b_3X_3+b_{12}X_1X_2+b_{13}X_1X_3+b_{23}X_2X_3+b_{123}X_1X_2X_3$$

Step 5: The matrix developed were solved using MATLAB for each of the property and the coefficient for the each of the property were calculated. Finally regression equations for each property were obtained.

Step 6: All regression coefficients and interaction coefficients were checked for significance using student's t-test.

3. Experimental Work

3.1 Raw material analysis: The raw material Lantana Camara in the form of chips was ground in jaw grinder and screened with 80 and 200 mesh sieves. The dust fraction which passed through 80 mesh and retained on 200 mesh screen was taken for analysis. The composition of the Lantana camara is given below:

Table 1: Raw material analysis

S.No.	Parameters of proximate analysis	%
1	Moisture Content	8.352%
2	Cold Water solubility	13.26%
3	Hot Water solubility	17.44%
4	1% NaOH solubility	26.71%
5	Weight. Reduction after Benzene Alcohol Extraction	8.461%
7	Hollocellulose	75.01%
8	Lignin	18.21%
9	Pentosans	17.02%
10	Silica content	2.31%

3.2 Pulping

Pulping were carried out in Weverk digester which is provided in four bombs .As mentioned above the cooking experiments were carried out according to design of experiments. In each pulping bath ratio was kept constant at 4:1 and active alkali was also constant i.e. 17%.The results of cooking experiments is tabulated below.

Table 2: Experimental Results

Time' hr	Temperature ^0C	Sulphidity %	Screen Yield %	Reject %	Overall Yield %	Kappa number	Brightness % ISO
1.5	150	15	26.28	30	56.28	33.18	19.2
1.5	150	17	29.34	25.7	55.04	28.84	20.9
1.5	150	19	32.76	22.38	55.14	26.3	22.1
1.5	160	15	28.55	24.32	52.87	31.88	21.8
1.5	160	17	30.43	20.28	50.71	27.65	22.1
1.5	160	19	33.52	22.02	55.54	25.16	26.3
1.5	170	15	29.22	23.58	52.80	26.54	22.1
1.5	170	17	31.63	20.28	51.91	25.52	22.6
1.5	170	19	34.21	18.04	52.25	24.3	28.3
2	150	15	26.69	31.62	58.31	31.53	20.9
2	150	17	33.25	16.62	49.87	25.33	26.1
2	150	19	33.88	16.9	50.78	25.3	26.1
2	160	15	26.98	25.49	52.47	30.69	22.9
2	160	17	34.54	15.62	50.16	24.96	26.3
2	160	19	34.84	17.37	52.21	24.3	28.9
2	170	15	29.64	15.36	45.00	29.3	24.8
2	170	17	34.84	15.37	50.21	24.4	28.9
2	170	19	35.66	11.5	47.16	21.87	30.5
2.5	150	15	27.22	17.89	45.11	29.21	23.5
2.5	150	17	29.32	20.56	49.88	26.33	26

2.5	150	19	35.96	9.35	45.31	23.4	24.8
2.5	160	15	27.39	18.33	45.72	29.08	23.8
2.5	160	17	31.04	15.34	46.38	25.08	26.3
2.5	160	19	36.83	7.81	44.64	22.8	26.8
2.5	170	15	28.39	16.3	44.69	28.04	24.2
2.5	170	17	31.25	16.54	47.79	24.03	30.4
2.5	170	19	36.98	8.04	45.02	21.07	28.4

4 Results And Discussion

4.1 Effect of pulping variables:

As indicated the variables are time, temperature, sulphidity which affect the yield, kappa number. brightness, lignin content . The effects of all these variables are shown below.

4.1.1 Effect of temperature on screened Yield

It is evident from the graph, the screen yield increases with increase in Sulphidity. It is due to the reason that the sulfide in cooking liquor greatly accelerates delignification and also higher Sulphidity helps to prevent the loss of cellulose during cooking. The effect of temperature on screened yield is shown in figure 4.1, 4.2 and 4.3 for 1.5, 2.0, and 2.5 hours cook respectively.

Variable	Range	Interval
Cooking temperature,°C	150-170	10
Time, hr	1.5-2.5	0.5
Sulphidity ,%	15-19	2

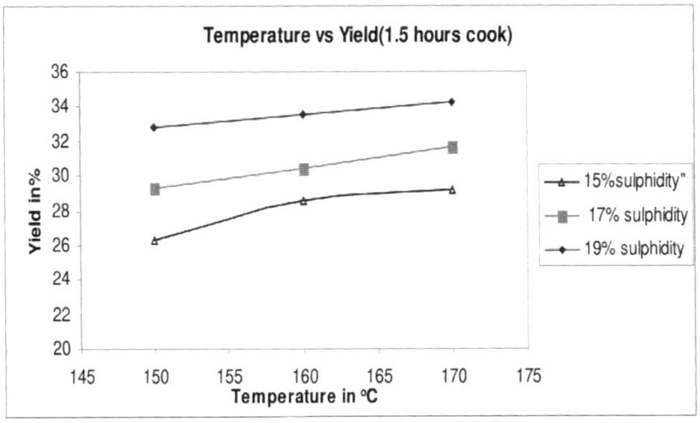

Fig 4.1: Graph showing relation between temperature and yield after 1.5 hour cooking at different sulphidity

In this case of fig 4.2 (2.0 hours cook) graph for 17% and 19% sulphidity coincides with each other and is an exceptional case. It may be due to experimental error .

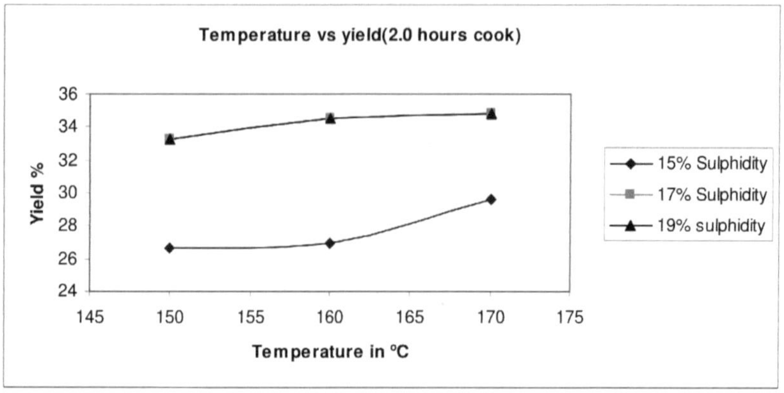

Fig 4.2: Graph showing relation between temperature and yield after 2.0 hour cooking at different sulphidity

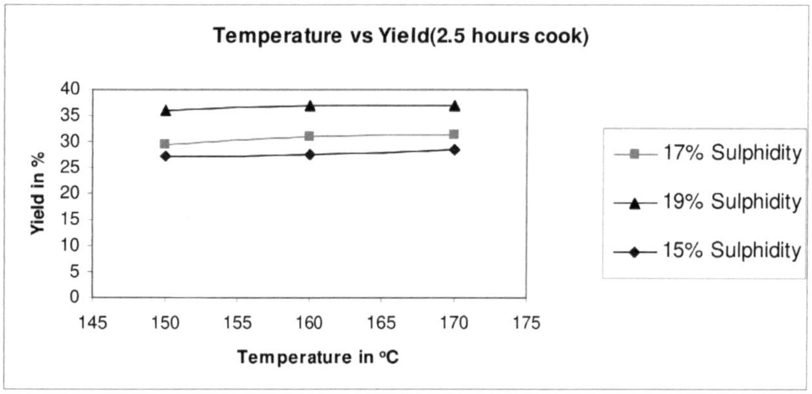

Fig 4.3: Graph showing relation between temperature and yield after 2.5 hour cooking at different sulphidity

4.1.2 Effect of temperature on brightness

It is evident from the graph that increase in sulphidity (at constant cooking time) increases the brightness. This may be due to the reason that sulphide in cooking liquor protects the carbohydrate fraction from the attack of OH^- ions. The effect of temperature on screened yield is shown in figure 4.4, 4.5 and 4.6 for 1.5, 2.0, and 2.5 hours cook respectively.

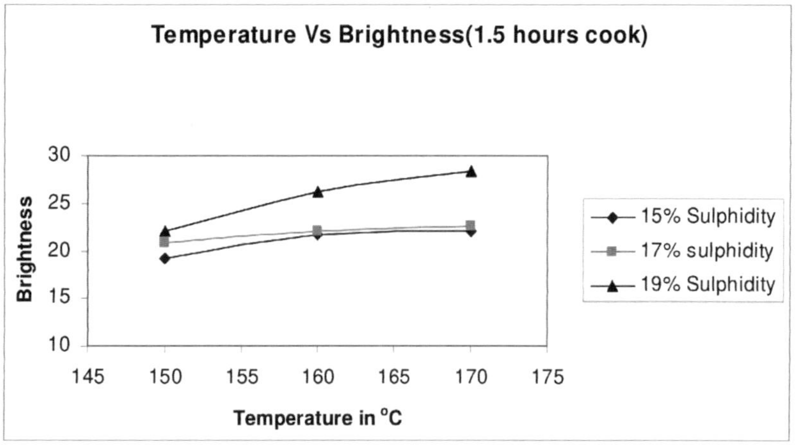

Fig 4.4: Graph showing relation between temperature and brightness after 1.5 hour cooking at different sulphidity

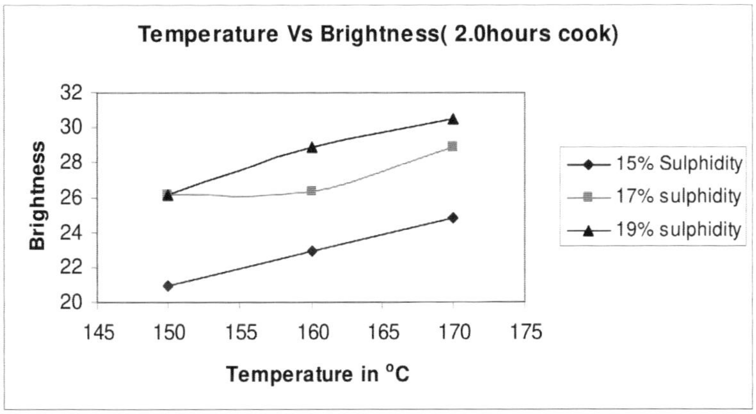

Fig 4.5: Graph showing relation between temperature and brightness after 2.0 hour cooking at different sulphidity

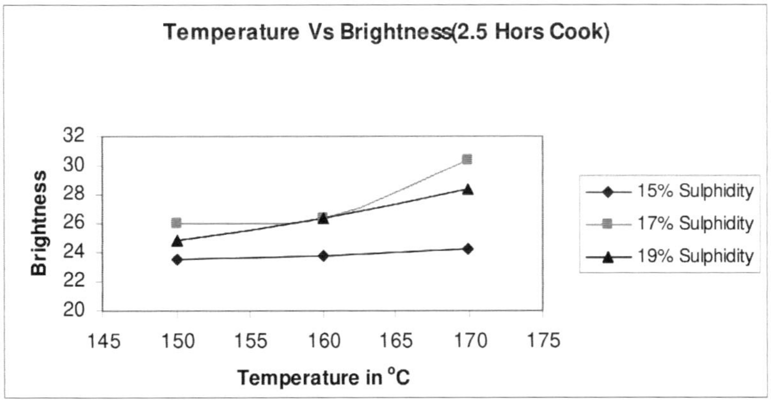

Fig 4.6: Graph showing relation between temperature and brightness after 2.5 hour cooking at different sulphidity

4.1.3 Effect of temperature on kappa number

An increase in sulphidity will decrease the kappa number. This is due to the reason that sulphidity promotes the delignification rate. . The effect of temperature on screened yield is shown in figure 4.7, 4.8 and 4.9 for 1.5, 2.0, and 2.5 hours cook respectively.

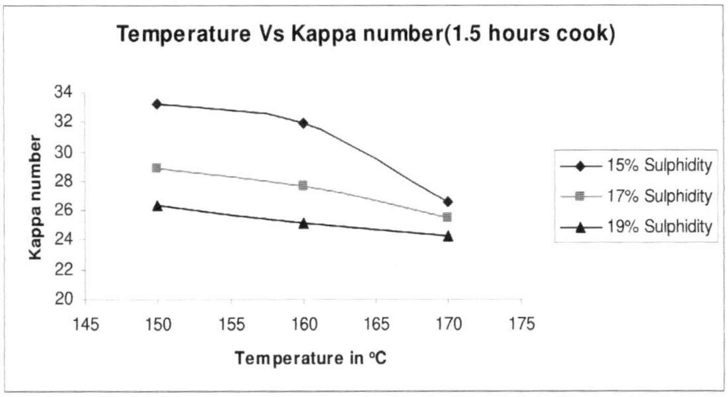

Fig 4.7: Graph showing relation between temperature and kappa number after 1.5 hour cooking at different sulphidity

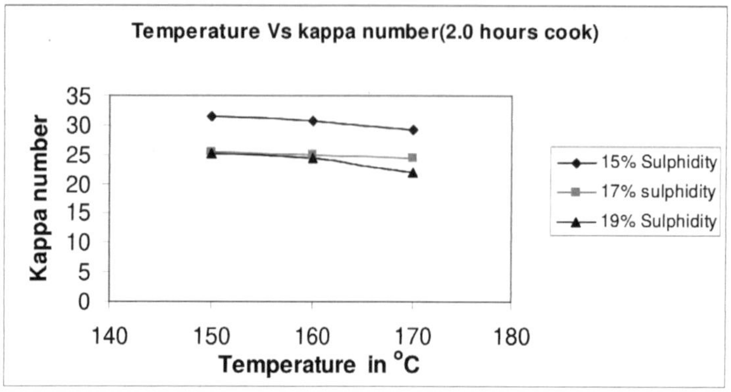

Fig 4.8: Graph showing relation between temperature and kappa number after 2.0 hour cooking at different sulphidity

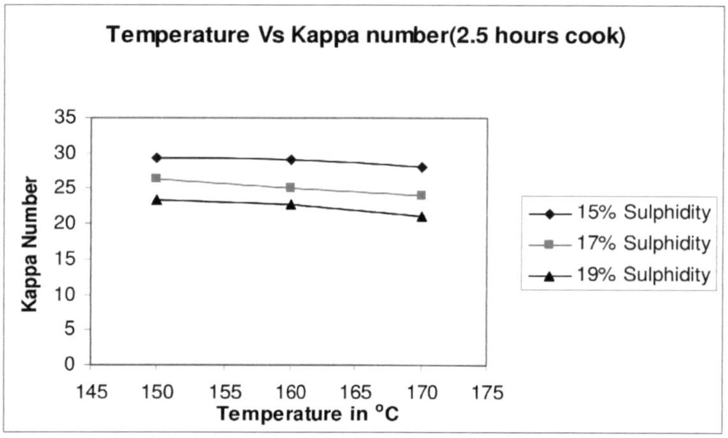

Fig 4.9: Graph showing relation between temperature and kappa number after 2.5 hour cooking at different sulphidity

4.2 Regression equation

Based on the experiments done data were generated. The following regression equation were obtained on the basis of experimental data collected applying full-factorial design with three levels and three factor after significance test of all coefficients through students t test.

For Yield: $Y_{Yield}=31.690+0.6733X_2+3.5711X_3$

For kappa number: $Y_{Kappa}=26.520-1.1272X_1-1.355X_2-3.0506X_3$

For Brightness: $Y_{Brightness}=23.33+0.3558X_1-0.767X_2+4.592X_3$

For lignin Content: $Y_{lignin}=9.3681-0.5306X_1-2.9767X_3+3.0506X_{23}$

Here: X_1=Time in hours

X_2=Temperature in ^{O}C

X_3=Sulphidity in %

6. Conclusion

Pulping of Lantana camara plant was carried out at different temperatures, time and sulphidity to develop a linear, multivariable statistical model. Multivariable regressional equation is developed to examine the effect of variables on pulp properties. From raw material analysis and pulpings the following conclusions were made.

- Best pulping result was obtained at 2.0 hours, 170^0C and 19% sulphidity. At these conditions the yield was about 35.66%, reject was 11.5%, overall yield was 47.16% and brightness was highest at 30.5.
- The screen yield was in the range of 26.28% to 36.98% which is less as compared to other non wood based raw materials.
- Even though pulpings at different conditions of temperature, time and sulphidity level were done, the reject percent could not be minimized, even with a high H factor. The reject percent was high varying from 7.81 % to 30 %.
- The percentage of extractives in Lantana camara was high and creating foaming problem during disintegration and washing. This may become a serious problem during production at large scale.
- Regression equations were developed for different properties of pulp like kappa number, yield, brightness, lignin content. These equations are

  For Yield: $Y_{Yield}=31.690+0.6733X_2+3.5711X_3$

  For kappa number: $Y_{Kappa}=26.520-1.1272X_1-1.355X_2-3.0506X_3$

  For Brightness: $Y_{Brightness}=23.33+0.3558X_1-0.767X_2+4.592X_3$

  For lignin Content: $Y_{lignin}=9.3681-0.5306X_1-2.9767X_3+3.0506X_{23}$

Here: X_1=Time in hours

X_2=Temperature in ^OC

X_3=Sulphidity in %

More pulping studies with other cooking processes soda, oxygen alkali and other pulping process are necessary to evaluate the potential of Lantana Camara for paper making. Excessive foaming during disintegration and washing can create serious problems at industrial scale and therefore need to be taken care of.

References

1. http://www.issg.org/database/species/ecology.asp
2. http://www.sbs.utexas.edu/bro406d/images/Lantana camara
3.http://www.hear.org/starr/hiplants/thumbnails/htm/lantana camara
4. http://weads crc.org.au/ documents/ wmg-lantana-pdf
5.http://www.fs.fed.us/global/iitf/pdf/shrub/lantana/%20camara pdf
6. http:/fleppe.org/id-book/Lantana %20 camara.pdf
7. Drapper, N. And smith, H. "Applied Regression Analysis" Second addition John Wiley and Sons, New York, (1966)

533f Enhancement Of Cellulose Sulfation Process Using Ionic Liquid Pre-Treatment Step

Zhaomei Wang, Lin Li, and Siyuan Guo. Research Institute of Light Industry and Chemical Engineering, South China University of Technology, No.381, Wushan Road, Guangzhou City, China

Abstract. Cellulose is the earth's most widespread renewable resource occurring at a volume of 700 billion tons. Value-promoting utilization of cellulose has aroused extensive attention throughout the world. Sulfation modification contributed to transferring cellulose into bio-active form cellulose sulphate (CS), which showed anticoagulant, antibacterial and antiviral activities. Increasing the degree of sulfation (DS) and improving the distribution of sulphate group along cellulose main chain are crucial for the activity. The highly ordered structure of cellulose has to be overcome in order to make the sulfation process viable for producing CS with expected bio-activity. In this study, an attempt was made to disrupt the cellulose structure using the ionic liquid, 1-butyl-3-methylimidazolium chloride ([C4mim] Cl), in a cellulose regeneration strategy which improved the subsequent sulfation reaction. Pretreatment process includes dissolving of cellulose with [C4mim]Cl and then following by regenerating with anti-solvents. De-ionized water, ethanol, actone and DMF were regarded as possible anti-solvents. The reaction parameters, including molar ratio of $ClSO_3H$-DMF/anhydroglucose units(AGU) in cellulose, the effect of temperature and duration of the reaction were considered. The final products CS were characterized by UV,IR, ^{13}C-NMR,GPC spectroscopy and elemental analysis. Results indicated that DMF appeared to be the most suitable anti-solvent according to the analysis of Sulfur content and intrinsic viscosity of CS. Pretreatment was found to increase the DS of CS from 0.86 to 2.13. A totally C_2 and a partially C_3 sulfate substitution were observed in CS derived from pretreated cellulose. It demonstrated that pre-treatment with [C4mim]Cl contributed to the uniform distribution of sulphate group along cellulose chain. The degradation of cellulose during sulfation process was delayed till 4h after reaction with pretreated cellulose as raw material, in contrast to only 1h after reaction with un-pretreated cellulose. The possible mechanism for enhancement of cellulose sulfation was due to the disruption of highly crystalline of cellulose ,based on the comparison of the structure of regenerated and unpretreated cellulose.

Can Ubiquitous Computing Be Cheap?

Dr. Paul B. Rasband, Chief Scientist, Vue Technology, 103 North Pointe, Lake Forest, CA 92630

Technology Lifecycles It's been forty-five years since Everett Rogers introduced the famous curve so often used to describe how technologies move through the various phases of adoption: (1) innovation, (2) early adoption, (3) early majority, (4) late majority, and (5) laggard adoption (see Figure 1). Obviously this is a simplification of the realities encountered by most new technologies, especially those which require co-adoption of other new technologies, business processes, and social attitudes.

Figure 1 - Rogers Technology Adoption Model

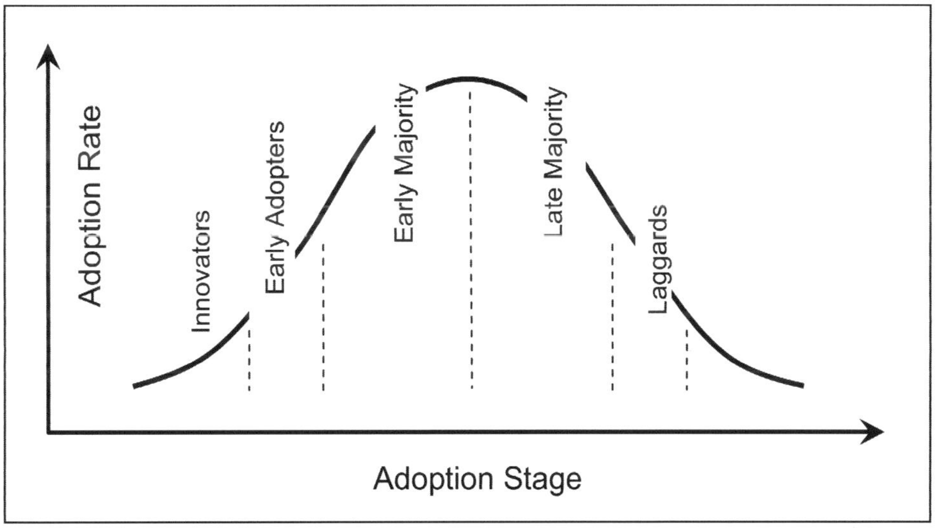

Some technologies are adopted slowly and poorly because they rely upon enabling technologies which are themselves under-developed and unevenly adopted. Later, when the enabling technology is replaced by a better approach is more widely embraced and more effective at laying the groundwork, the technologies relying up it can find new life. When this layering of technologies goes many layers deep the situation can become truly complex, and might better be modeled as a multi-dimensional hypersurface. Because "ubiquitous computing" is really a collection of many related applications and enabling technologies, we might expect its adoption to be more complex than most "stand alone" technologies.

A big catch-22 faced by the vendors of radically new products is that, even though many innovators are willing to pay a great deal for the new product, most early adopters are more price-conscious. And, because of the low volumes the vendors might have a hard time reducing costs rapidly enough to keep the market moving. Prices can't be reduced until the market grows, and the market can't grow without big price reductions. Some of described this situation as a "chasm" between innovators and early adopters, and neglect of it has destroyed many new products and enterprises.

There are many new "toy technologies" which capture the interest of small groups of people, sometimes for many years, while the vendors of those new technologies search for the "killer application" which will propel the technology into the mainstream and pull their companies into hyper growth mode. Together the word processor and the spread sheet were the applications that made the personal computer a ubiquitous tool. Vendors of almost any new video display technology dream of finding a way to make the new technology attractive to television designers - that's their killer app. GPS was cool for police and hikers, but when its use in rental cars made the general populous aware of the huge utility of real-time positioning, that market took off and is now rapidly moving into early majority status. One reason to search for the killer app is that it makes the chasm smaller and less formidable (mostly because it gives vendors, venture capitalists, banks, the stock market, and other financial stakeholders enough courage to actually subsidize the crossing of the chasm by investing in excess production capacity).

The real trick for new business trying to predict and traverse the adoption curve in the real world, especially the world of ubiquitous computing and its enabling technologies (PDAs, WiFi, OLED displays, ad hoc sensor networks, printed electronics, RFID, smart packaging, and on and on) is that the curve is in fact a very complex hypersurface with overlapping and interweaving technical, economic, and social issues. This, according to most opinions, should slow adoption (even severely) and that makes predicting the chasm (or chasms) even more critical for those entrepreneurs who don't want to see their businesses waste capital and starve to death. The faith of the entrepreneurs is that there is some manageable route to large volumes for carefully chosen and bundled products in this space. This paper deals with elements of this complex adoption hypersurface which are key to developing the right products in the right order for the right markets.

Looking Forward To That World With Ubiquitous Computing Are there any killer apps waiting out there for ubiquitous computing? That depends in part how we define "computing", so let's start there. Automated computing became much more ubiquitous when it made the migration from the environmentally controlled computer room to the desktop. In the 1990's it made another huge step forward, jumping into handheld devices. Is there another jump to make?

One intriguing possibility involves the **re-usable newspaper**. This would involve some form of flexible (even rollable) display interfaced with a memory chip, small power supply, and control unit. For data input it might involve a hard port (USB or the like), or a transceiver chip and antenna for wireless input. The basic idea is that the user would upload the news into the device each day (or several times a day) and carry it with them in a convenient form in the cab, train, coffee shop, lobby, waiting room, or conference room (before the meeting starts). The upload might be consciously triggered, or it might be automatic (say, when the user encountered a WiFi hotspot). Some obvious challenges involved with the development of the re-usable newspaper include pure hardware issues, and are discussed in more detail below. However, even if the ideal piece of hardware were available today at a good price, there would remain the separate challenges of support infrastructure and content supply. Someone needs to supply the wired and/or wireless network, provide the news in the proper format, and create the business relationships to make sure everyone gets paid for their efforts. The evolution of this environment takes time.

Another potentially popular ubiquitous computing application that has been discussed in recent years would use **video displays printed onto consumer packaging** (like a package holding a computer game disk, or a power tool, or even a kids' cereal) to advertise the product more effectively on the retail shelf, or to provide improved user instructions.

Another example of an interesting application is **portable (stowable) solar panels** for consumer electronics. This application, from the perspective of enabling technology, has a lot in common with the re-usable newspaper. In the conventional vision of this product the solar cells are printed on a flexible substrate and folded out or unrolled when power is needed for a computer, PDA, iPod, or other such device (in the park, at the beach, and so on). If the technology is up to the task, it might even be mainstreamed and used to provide power for a home at a cost much lower than is possible with today's solar cell products.

As a final example of a ubiquitous computing application, we will consider a much more complex "product" than those mentioned above - something we might call a **senior-support home environment**. Like many ubiquitous computing applications this actually involves a combination of simpler technologies working together to create an advanced set of related solutions. One way of describing this application would be VMI (vendor-managed inventory) of food and drugs for senior citizens. The idea is that since seniors are often home-bound and consistently use the same products, they would be ideal for a subscription service in which food and drug retailers (or, perhaps, even consumer product manufacturers in some cases) monitor the inventory of products in the home and re-supply the senior citizen on a timely basis, with a minimum of initiative or attention required on the part of the senior. This would require (in one scenario) the RFID-tagging of consumer product packaging, and the installation of an RFID reader infrastructure in critical points in the home (kitchen cabinetry, refrigerator, medicine chest, etc). Related applications have been discussed involving drug regimen compliance support for seniors - a service where the proper use of prescription drug products by seniors is monitored remotely (again using RFID combined with web-based software). Put these together with remote fire and heat monitoring, emergency assistance requests, and general activity monitoring, and you get a total package which amounts to comprehensive remote support for seniors' safety and quality of life. This solution package would require a rather sophisticated home network interfaced with some web services and hosted software applications, not to mention some creative business relationship development, and some business ecosystem development to provide installation and support services. And, not inconsequentially, it would also require (practically) the manufacturer-supplied item-level RFID tagging of a critical mass of food and drug products - something which is only now beginning to occur, and will take years to evolve to an extent sufficient to support the solution as described above.

Usage Drivers Behind Ubiquitous Computing The applications just described are clearly just a few examples of many that might be proposed. In fact, in the real world today the definitions of "ubiquitous computing", "smart homes", "smart packaging", "smart spaces", indeed smart-anything are being blurred as dozens of similar applications are being pushed along various stages of technical and business development. What is driving these new technologies? Perhaps a better question which the potential investors considering any given technology must ask themselves is, "*Are there* sufficient and compelling drivers for this technology?" A killer application will offer a significant fraction of people on planet at least one of the following:

Convenience There seems to be a never-ending quest on the part of people who want information and answers right now. When they are faced with a question, they want it answered immediately. Many of the ubiquitous computing applications attempt to predict what a user will want to know under specific circumstances, identify those circumstances, get the attention of the user in those circumstances, and then pass the information to the user in an automated way.

Simplicity Some ubiquitous computing applications are designed to replace older applications which require more cumbersome gadgets and/or which require users to complete relatively complex tasks to get a job done. Today, any new technology which is not simple to use is almost surely doomed.

The Green Thing Eliminating waste, reducing power consumption (through reduced travel demand and consolidation of item transport), facilitation of recycling, and other environmentally motivated interests may be satisfied by various new ubiquitous computing applications. These are not just fringe benefits - they could drive many new technologies.

Novelty In a world where simple needs of food and shelter are easily met, the role of entertainment, novelty, fashion, and other "non-essential" desires become much more important. Novelty might not be a permanent driver, but it can be a hugely important force in the early stages of technology and business development when the new application is most vulnerable.

Security Many of the new ambient intelligence technologies are ideal for sensing dangerous situations and providing alarms or alarm triggers. Food safety, fire safety, intruder alerts, collision prediction, and other benefits are possible. Also, things like RFID and biometrics can be used to combat fraud and theft.

Productivity Businesses can benefit from increased simplicity and convenience (see above) in the lives of their employees. And since B-to-B pathways of new product introduction are often more efficient than mass marketing, many entrepreneurs are concentrating their distribution efforts on business customers.

Enabling Technologies Figure 2 shows some "hard" technologies that are being developed to enable ubiquitous computing applications of one sort or another. Two things need to be stated at the outset. First, the technologies listed in Figure 2 are the tip of the iceberg for hardware-oriented developments underway. Second, software approaches and the efficient layering of software standards are in many ways more difficult and more critical to the ultimate success of ubiquitous computing applications like those previously discussed.

Fundamental Technical Challenges In order to understand the challenges faced by research and product development teams working on these new technologies (and to appreciate their accomplishments to date, which are impressive), one can start by consider two simple principles: First, the time required by an electron (or electron hole) to travel a given distance in a circuit or device is determined by (a) the distances involved - that is, device **feature size**, (b) the characteristic speed or **mobility of the electron** (or hole), and (c) the driving potential or **voltage**. The second key principle is that the switching speed of a field-effect transistor depends largely upon how much charge must be moved, which in turn depends upon the size of the transistor itself. Thus, the pursuit of higher switching speeds for printed transistors is in large part a pursuit of (a) finer feature size potential for the printing processes and materials used (without compromising circuit integrity), and (b) higher mobility for electrons in conductors and semiconductors.

Electron and hole mobility is determined not only by the intrinsic electron-orbital-based properties of the pure semiconductor materials in perfect crystal structures, but also by impurities in the materials, defects in the crystal structure, and the range of order in that structure. For maximum mobility one desires very long range order (over hundreds of nanometers), and obtaining this with polymers heavy enough to effectively print has been a challenge for technologists. One can achieve very high order and high mobility in semiconductor films using vacuum-deposited small molecules, but this approach fights against the desire to move away from expensive and slow processing equipment. An upper limit for mobility in organic films is between 1 and 10 $cm^2/V.sec$ (Ref. 1). Defects and impurities reduce mobility below this ideal, sometimes by orders of magnitude. For comparison, consider that electron mobility in single-crystal Si at room temperature is 10,000 $cm^2/V.sec$.

Figure 2 - Areas of Technology Endeavor for Ubiquitous Computing

Materials Development	Process Development	Device Development
Conductive Inks	Litho, Gravure and InkJet Printing with the New Inks, and Fine Features	OLEDs
Organic Semiconductors	Micro-Contact Printing	Printed Transistors
Barrier Film Polymers	Web-Based Photolithography	Optical Switches and Interfaces
Substrates with proper ... - High-Temp properties - Smoothness / Printability - Purity - Flexibility - Mechanical Strength and Flex endurance - Cost (web producibility)	Hermetic Sealing	MEMS-based resonators, inductors, capacitors, switches
Conductive Adhesives	Super-fine Chip Singulation, Handling, and Attachment	Novel Sensors
Dielectric Adhesives	Film Printing and Curing	Printed Batteries
Colloidal Suspensions	Amorphous material film annealing (Si, small molecules, etc.)	
	High(er) Speed Screen Printing	

Another key issue in developing printed digital electronics is **power consumption**. For traditional single-crystal Si based integrated circuits, a key to reducing power consumption (and thus enabling true portability for ubiquitous computing gadgets) has been to reduce the transistor **switching voltage**. If devices can be run at lower voltage, they use less power, use smaller batteries, and run for longer periods of time. Of course, the circuits and the applications using them run more slowly when voltage drops, so the desire for low power consumption and high speed involves a tradeoff - and a tradeoff that is much more stringent in organic devices than is observed for single-crystal Si.

Rounding out a short list of the technical challenges facing printed electronics is the need for greater thickness in the printed conductors (to reduce circuit impedance), balanced by a need for short drying / curing times (again, requiring a tradeoff).

Because of these challenges for printed digital electronics - which are large but not necessarily fatally so - many technology strategists in this space have decided to keep their focus on single-crystal Si chips, and use traditional microlithography combined with new approaches to cutting, handling, and attaching extremely small chips on (or in) very non-traditional substrates to produce very low cost gadgets with high speed and low power consumption. Mechanical handling of chips as small as a few hundred microns on an edge is obviously challenging - perhaps as challenging as solving the alternative problems associated with printed digital logic approaches.

Social and Cultural Challenges The technical challenges just described almost surely have solutions which will be found given time and money. Generally, we under-estimate the abilities of research and development teams to solve hardcore technology challenges. However, it should also be noted that we just as frequently *over*-estimate the ability of new businesses and product development teams to solve the problems associated with user psychology, value perception, and sociological acceptance of the products being developed.

Part of the difficulty is that many of the problems addressed by ubiquitous computing and ambient intelligence solutions simply require more flexibility than a single gadget or software product, or even a set of products, can provide. Consider the re-usable newspaper again. People buy a newspaper for the news, of course, but they also buy it to clip coupons, line the bird cage, or give the neighbor kid a morning job delivering the paper. There are also even more intangible preferences that newspaper readers themselves might not be conscious of - like sharing the paper with their spouse at breakfast (e.g., you take the financials while I look at the sports section). These issues probably don't make it into the business plan of the Acme e-Newspaper Co., but they might be very important to the ultimate success of the company.

Economic Challenges Sometimes new technologies fail (or have a slow start) simply because they require the cooperation of companies in separate industries with very different economic climates, risk tolerance, and patience of uncertainty. This is especially true, perhaps, in the budding field of printed electronics. For example, the manager of a specialty pharma packaging division might be excited about a series of digital electronics printing trials as long as his or her biggest packaging customer is interested in the project. But next year, if the project feels less urgent and he or she is asked to pitch in $100,000 of the development cost, enthusiasm may wane. After all, they must live in the real world.

Migration Pathways To Ubiquitous Computing There are not many new ventures trying to develop smart wallpaper or expert-system-equipped trash bags. But there are development projects and new ventures launched which are aimed at a few less spectacular applications, mostly centered in flexible displays and RFID. The interest in RFID stems from a conviction on the part of entrepreneurs that "traditional" RFID tags based on flip-chip assembly of silicon chips on etched or stamped antennas cannot produce (even in volumes of trillions of tags per year) a market tag price less than about 5 cents. If printed digital logic using organic semiconductors or even amorphous silicon thin films can reach 1-2 cents without sacrificing performance too much, then the future of mass-market passive RFID may not lie with single crystal silicon, flip-chip assembly, and the traditional approach. One rather large application space which may be even closer for printed digital logic (mostly because the demands on read distance and read speed are less severe) involves smart cards (ID cards, bus and train tokens, stored value / gift cards, etc).

The role of printed circuits in RFID reader antenna infrastructure is probably less widely appreciated at the current time, but has great potential. For example, printed antenna traces in the inner layers of laminated counter-tops is a real possibility (Ref. 2). As item-level tagging begins to penetrate into consumer products and becomes more of an everyday reality, there is no doubt that entrepreneurs (in garages and large companies) will start to concentrate more on "smart kitchen" and "smart bathroom" technology, and that will lead to some interest (if turbulent) interaction between electronics engineers and building material company managers and engineers.

The challenge of plotting a migration path through the space of ubiquitous computing really involves looking for a real value proposition in a real market, close enough to imagine capturing, and then finding the minimum-risk path to a complete solution in that market. This includes not only technology development, but mass psychological analysis, company culture matching, and investor expectation management.

References

1. Pope, M, and Swenberg, C. E., Electronic Processes in Organic Crystals and Polymers, 2nd Edition, Oxford University Press, New York, 1999, pp. 337-340.

2. U.S. Patent 7,209,039, Krebs, et al, Illinois Tool Works, April 24, 2007.

607b Development of An Rfid Lab to Evaluate Smart Forest Products

Gary F. Madsen, Auto-ID Sensing Technologies, Kimberly-Clark Corp., 2001 Marathon Ave., Neenah, WI 54956

Implementation of RFID poses numerous challenges that demand rigorous testing in real-world environments. Here we describe the challenges and successes experienced in developing Kimberly–Clark's accredited RFID laboratory. We will discuss some of the experiences around startup and maintenance of the lab, and share learnings around the technology and applications it to processes at Kimberly-Clark to improve current practices.

Printed Electronics at Western Michigan University

Paul D. Fleming III, Bradley Bazuin, Marian Rebros, Erika Hrehorova, Margaret K. Joyce, Alexandra Pekarovicova, and Valery Bliznyuk

Center for Ink and Printability Research
Center for Coating Development
Department of Paper Engineering, Chemical Engineering and Imaging
Western Michigan University
Kalamazoo, MI 49008

Abstract

Recent research in printed electronics from our group is reviewed. Functional materials are being developed and optimized for various printing methods, such as ink-jet, gravure and flexographic printing. In particular, flexography and rotogravure printing have been employed to print RFID antennae using metallic inks. Other materials under study include conductive, semiconductive, dielectric and resistive materials for printing integrated circuits used in RFID chips on paper and paperboard substrates. Effects of printing process parameters and in depth study of paper properties are in the progress. New testing procedures and application of known testing methods are used to predict ink performance during printing, ink-paper interactions and final performance. Antenna and tag performance have been extensively characterized in our RFID testing laboratory.

Introduction

Printing is increasingly seen as a means for the fabrication of advanced functional materials, such as electronic components. These include RFID tags[1,2], Displays[3,4], solar panels[5] and others. Printing of such devices requires conductive materials[1,2,6,7,8,9], semiconductive materials[10] and dielectric materials[10]. Conducting materials (inks) may be metallic based, such as required for RFID antennae[1,2], or conducting polymers[6,7,8,9,11], as required for conducting traces and source and drain electrodes in an integrated circuit. Semiconducting materials may be based on semiconducting polymers[10,11,12], or small organic molecules and crystals[13]. Here, we will concentrate on inorganic metallic conducting materials, particularly silver, and conducting polymers, where our recent research has been conducted.

Lowering the cost of RFID tags is nowadays a big driving force for innovations in printable materials and associated devices. An RFID tag consists of an antenna and an integrated circuit. Currently, only 10% of RFID tag antennae are printed, typically using silver based inks [14,15,16]. These materials may be screen-printed, flexographically printed, gravure printed, offset printed or digitally printed on a variety of substrates ranging from paper, Mylar polyester, polycarbonate, polyimide, PVC, to glass. Most silver-based materials are based on special "leafing" silver flakes [17] that are designed to float to the surface during drying to promote high surface conductivity. The silver-based materials are well suited for printing RFID [18,19] antennae and circuit interconnects, but are not appropriate for printing of contact electrodes in integrated circuit (IC) components, due to the large size of silver flakes, which however, can be overcome by employing nanosized metallic particles[15,16]. Besides the metal filled conductive materials, solution processable conductive and semiconductive polymers are already leading to low cost sensors and cheap, disposable electronics[11]. Our work to date has focused on two conducting polymers, polyaniline (PANI)[6] and poly (3,4 ethylenedioxythiophene)-poly(styrenesulfonate) (PEDOT:PSS)[7,8,9]. PEDOT:PSS based inks have been used to print conductive layers[9]. This conductive polymer is commercially available as a water-soluble polyelectrolyte system with good film-forming properties, high visible light transmittance, and excellent stability[20]. Some applications of PEDOT:PSS include antistatic coatings, conductive layers in organic light emitting diodes (OLEDs), capacitors and thin film transistors[21]. It has been reported that electrical conductivity of PEDOT:PSS can be enhanced by the addition of different organic compounds[22]. The conductivity improvement is strongly dependent on the chemical structure of the compound. Among the alcohols, ethylene glycol and glycerol were found to be the most efficient. Enhancement of conductivity is believed to be a result of an increased inter-chain interaction caused by conformational change of the PEDOT chains from the coil structure into expanded-coil or linear structures[23].

ICs typically consist of passive and active building blocks and require at least three main material properties for construction, such as conductivity, semiconductivity and dielectric (insulating) properties. Optimal performance of the ICs can be obtained if materials with a wide range of band gaps and conduction mechanisms (e.g. p or n) are available. Considering manufacture processes, transistors using silicon require a complicated and high precision manufacturing process that creates electronic circuits by starting with a single-crystal silicon substrate and applying a variety of processes such as oxidation, the addition of impurities (ion implantation), chemical-vapor deposition, thermal diffusion, annealing, metallization, photolithography and etching[24]. Large scale and expensive equipment, such as clean rooms and vacuum systems, are required. In contrast, with printed transistors, one can take advantage of the features of organic materials and, by dissolving them in a solvent, use printing technologies such as rotary press or inkjet printing to create circuits simply[10,25]. Since these are low-temperature processes, they have excellent compatibility with plastic and paper substrates[26].

With the technology of flexible electronics becoming closer to device prototyping and commercial production, it is clear that the choice of substrates with desirable properties is essential in order to make this technology viable. Flexible substrates pose a number of challenges. Dimensional stability of the substrate is very important in order to ensure precise registration and resolution. Many types of substrates are also incompatible with some solvents used for organic materials[4]. Surface smoothness and cleanliness of the flexible substrate are both essential to ensure the integrity of subsequent layers and formation of a high quality interface for better device performance. Of the flexible plastic substrates, the most commonly used are polyesters[27,28] (PET, PEN) and polyimides[29]. Although paper is of big interest for printed electronics, there are very few reports to date[1,2,30].

Printed RFID Antennae

The primary focus of our project to date was to evaluate the performance of RFID antenna printed directly on packaging substrate, such as paperboard and label stock papers. RFID antennae were manufactured by two different printing processes (flexography and gravure) using silver filled conductive inks. The print pattern was designed with intention to evaluate electrical performance, as well as printability, of the two conductive inks primarily by using various features, such as conductive traces at different tone values and angles to the print direction, positive and negative line blocks and finally two different UHF RFID antennae designs (Figure 1). Three bands have been deployed to sample either different anilox roll line screens or different gravure cylinder engraving resolutions to find the most suitable image carrier parameters for desirable properties of the final prints.

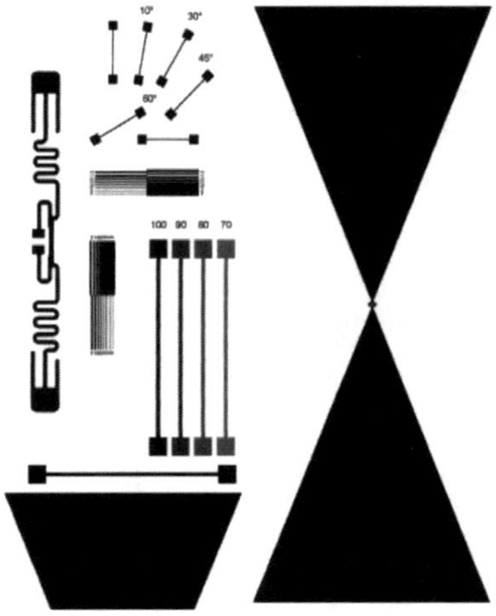

Figure 1 Illustration of the main design features included in testing print form.

Printing trials utilized 3 different silver-based inks based on difference chemistries, a solvent-based, a UV-cured and an aqueous-based. Seven different substrates were printed, consisting of 4 packaging papers and 3 packaging paperboards. In order to correlate the properties of materials used for printability and electrical performance, conductive inks and packaging substrates were fully characterized using standard methods. The summary of measured properties and methods used is given in the Table 1. Electrical characterization of printed features is given in more detail later in the article.

The definition of the printed edge is very important, in particular for RF performance. Therefore, the printed samples were analyzed for line raggedness[31]. An example of line raggedness measurement is shown in Figure 2. Raggedness is determined by the displacement of the black-white boundary line from the ideal boundary line. The ideal boundary line is determined by calculating the best-fit line through the boundary points.

Table 1 Measured properties and methods used		
Substrate characterization	**Ink Characterization**	**Print Characterization**
Roughness (PPS, Stylus Profilometry)	Rhelogy (Rotational Viscometry, Oscillatory Testing)	Line Width/Length (Image Analysis)
Porosity (PPS, Mercury Intrusion)	Surface Tension (Pendant Drop Analysis)	Raggedness (Image Analysis)
Pore Distribution (Mercury Intrusion, Gas Adsorption)	Particle Size Analysis (Dynamic Light Scaterring)	Ink Thickness (Optical Microscopy)
Topography (Surface Scan Analysis, AFM		Ink Film Roughness (Stylus Profilometry)
Liquid Penetration (Ultrasonic Transmission)		Electrical Resistance (AC/DC Multimeters)
Surface Energy (Contact Angle Analysis, Owens-Wendt Estimation Method)		RF Performance (RF Network Analyzer)

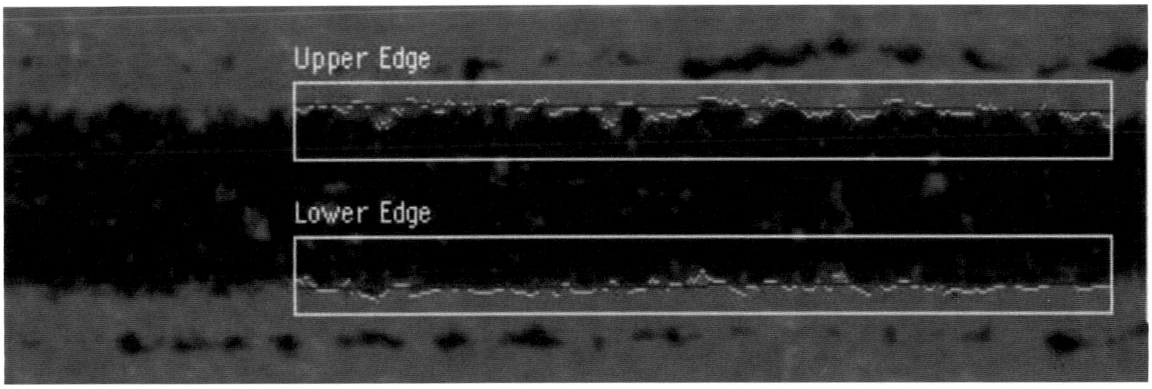

Figure 2 Example of line raggedness evaluation for a printed line.

Characterization of Electrical Performance

DC Resistivity

Determining material resistivity is accomplished by measuring the resistance of a piece of the material of known dimensions, such as a solid rod or bar or a thin line on a non-conducting substrate. As long as the material size is uniform and the cross-sectional area and length are known, the resistivity can be directly computed once the resistance is know. For this purpose, the printed line dimensions (length, width and thickness) need to be measured. The length and width of printed conductive traces was measured by image analysis and the thickness of the ink film was measured by optical microscopy. Samples for thickness measurements were embedded in a resin and then polished. Examples of an ink film cross sections are given in Figure 3.

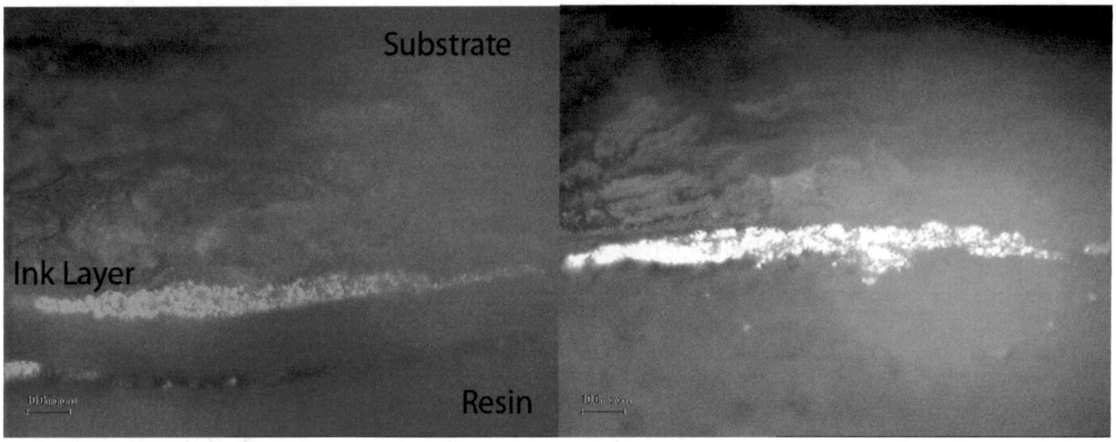

Figure 3 Illustration of ink film thickness for water-based ink 3 (left) and solvent-based ink 2 on a label paper

Figure 4 shows the results of bulk resistivity of flexographically printed samples with three ink systems on different packaging substrates. For better comparison of the different ink systems, the figure shows bulk resistivity values only up to 0.4 Ohm mil, because bulk resistivity of the ink system 3 was significantly higher than 0.4 Ohm mil. High resistivity of printed ink films resulted in poor antennae performance.

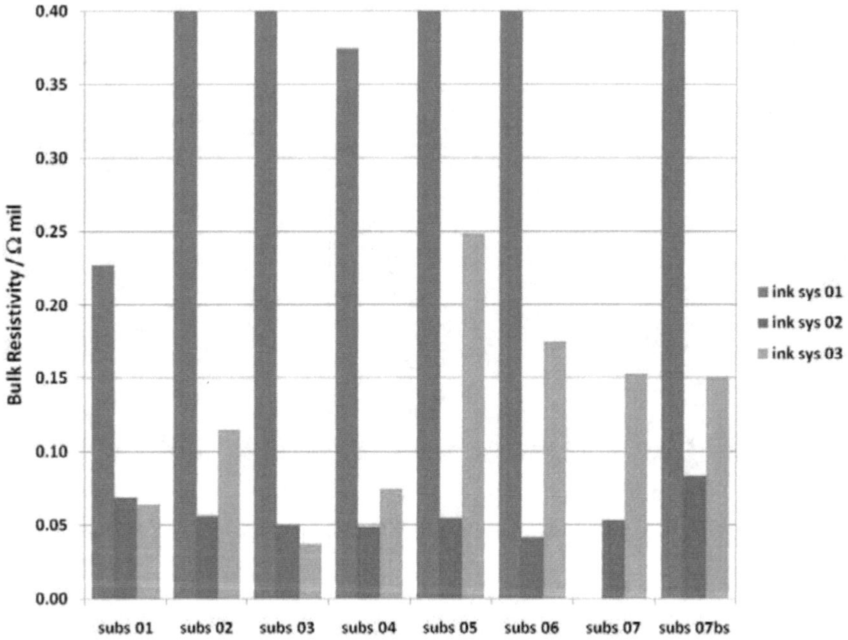

Figure 4 Bulk Resistivity

An accurate measure of resistance is obtained using a 4-point probe, where a current is caused to flow through the material using two probes at opposite ends and a voltage is measures at the desired length along the test resistor as shown in **Error! Reference source not found.**5.

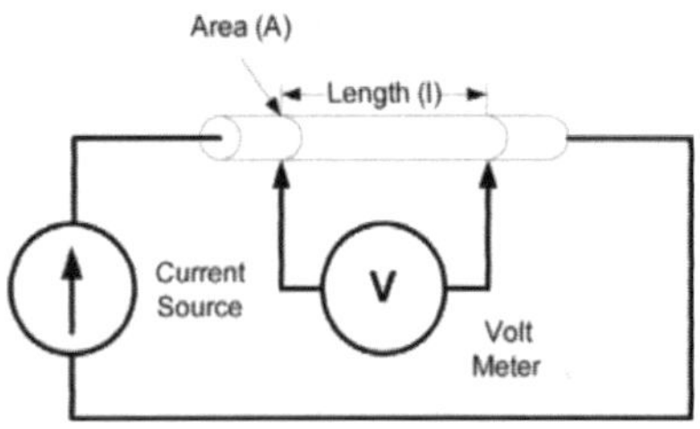

Figure 5 Four-point Resistance Measurement.

AC Impedance

A Milliohmmeter was used to perform AC impedance measurements. This instrument performs 4-point probe measurements using a 1 kHz current source reference. Using known current magnitude and phase, the device measures the magnitude and phase of the voltage waveform across the device under test (DUT) and computes the complex impedance in terms of

$$Z = R + jX \tag{1}$$

where R is the electrical resistance in Ω and
X is the reactance in Ω.

The sign of X defines whether the equivalent series impedance at 1 kHz is inductive (positive) or capacitive (negative) and can be used to estimate the magnitude of the inductance or capacitance.

The electrical resistance, R, measurement should also compare directly with the DC impedance measurement, thereby, providing verification and validation of the DC measurement (Figure 6)

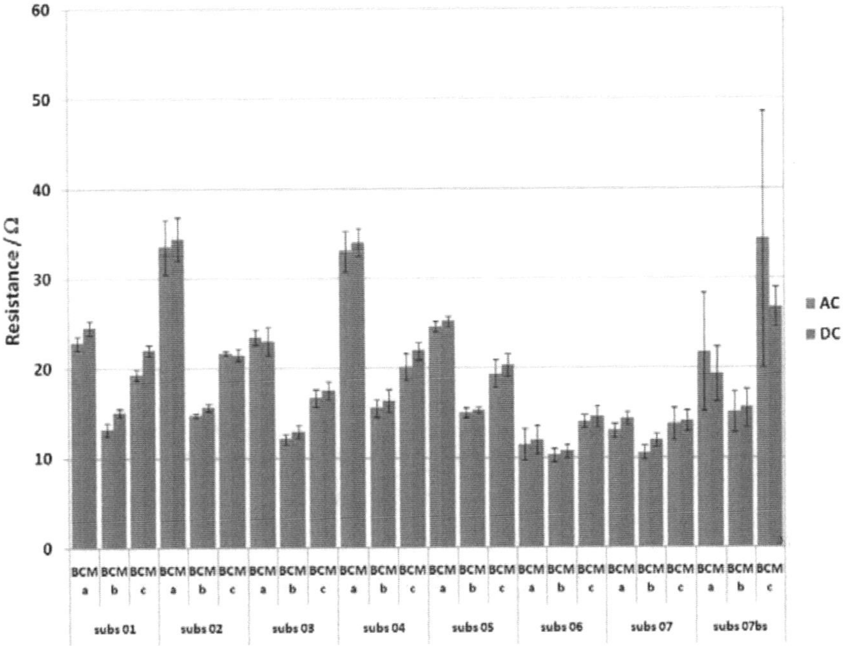

Figure 6 Comparison of AC vs. DC characteristics for measured substrates (ink system 02).

Impedance Measurement: RF

As signal frequencies increase in a conductor satisfying Ohm's Law, the electrical field and current density decays exponentially with the depth of the conductor [32]. The depth at which the values decay by 1/e of the surface values is defined as the RF skin depth.

$$\delta = \frac{1}{\sqrt{\frac{\omega\mu\sigma}{2}}} = \sqrt{\frac{2\rho}{\omega\mu}} \tag{2}$$

where μ is the material permeability in H/m, σ is the conductivity in S/m, and ρ is the material resistivity in Ω-m. and ω is the angular frequency in Hz.

As current does not flow throughout the entire conductor, the effective RF impedance of the material will be greater than the resistance value previously defined. A new cross-sectional area term based on the skin depth must be used as

$$R_{RF} = \frac{\rho l}{w\delta} = \frac{\rho}{\delta}\left(\frac{l}{w}\right) = \frac{1}{\sigma\delta}\left(\frac{l}{w}\right) \tag{3}$$

where w is the width of the material
l is the length.

Based on this equation, the surface resistance of a material is defined in terms of ohms per square (based on a square of material where $l = w$).

$$R_S = \frac{\rho}{\delta} = \rho\sqrt{\frac{\omega\mu}{2\rho}} = \sqrt{\frac{\omega\mu\rho}{2}} = \sqrt{\frac{\omega\mu}{2\sigma}} \quad (4)$$

Figure 5 shows the effective RF resistance as compared to the DC resistance previously defined.

If a circular conducting rod, as shown in Figure 5, is considered of diameter, d, the comparison between the DC resistance and RF resistance can be computed. First, comparing the total resistance

$$R = \frac{\rho l}{2\pi\left(\frac{d}{2}\right)^2} \text{ versus } R_{RF} \approx \frac{\rho l}{\pi d\delta} \quad (5)$$

where the entire circumference of the rod is assumed to be large in comparison to the skin depth for computing the RF area. Comparing these values,

$$\frac{R}{R_{RF}} \approx \frac{4\pi d\delta}{2\pi d^2} = \frac{\delta}{\frac{d}{2}} \quad (6)$$

The RF resistance is then

$$R_{RF} \approx R_{DC}\frac{0.5d}{\delta} = \frac{r}{\delta} \quad (7)$$

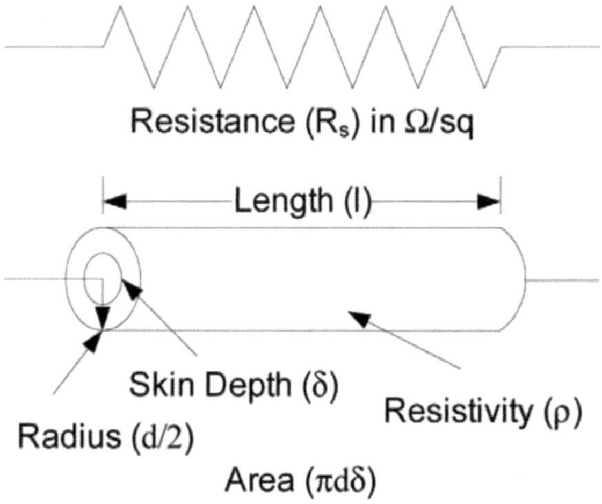

Figure 5 RF Surface Resistance

For a skin depth significantly less than the printed line height, the RF resistance is a significant multiple of the DC resistance!

To provide specific examples, the skin depth for selected metals and inks at 915 MHz, the DC resistance in Ω/sq/mil (25 um), and the surface resistance at 915 MHz is shown in Table 2.

Table 2 RF Skin Depth for Specific Materials at 915 MHz

Material	Resistivity ρ (Ω-nm)	Skin Depth δ (μm)	R_{DC} mΩ/Sq/mil	R_S mΩ/Sq	Reference
Silver	15.87	2.10	0.635	7.571	Resistivity [33]
Copper	16.78	2.16	0.671	7.785	Resistivity [33]
Gold	22.14	2.48	0.886	8.943	Resistivity [33]
Aluminum	26.50	2.71	1.060	9.784	Resistivity [33]
Ink 01	125.00	5.88	5.000	21.249	Ω/Sq/mil
Ink 02	150.00	6.44	6.000	23.277	Ω/Sq/mil
Ink-03	500.00	11.77	20.000	42.499	Ω/Sq/mil

Note (1): in the above table $R_{DC} = \frac{\rho}{(25\ \mu m)} = R\frac{w}{l}$ Error! Bookmark not defined..

Note (2): Resistivity for metals at 293° K [33].

Antenna testing has been performed using a continuous wave RF antenna tester. The initial RF testing was performed using a laboratory antenna measurement range. This testing is intended to provide both antenna beam pattern and comparative antenna performance measures for the various printed antennae provided.

The antenna range consists of a fixed reference antenna at a defined distance to a test antenna placed on a movable pedestal (360° horizontal movement and ± 45° vertical movement). A RF network analyzer is used to source a continuous wave RF signal at a known frequency and phase to the reference antenna. The antenna transmits the RF wave to the test antenna, where it is received as an input to the RF network analyzer. The network analyzer determines the incoming signals magnitude and relative phase as compared to the transmitted signal. The results are captured over an HP-IB network connection to a personal computer, where operational software controls the test and data collection. This measurement allows a three-dimensional pattern of the antenna performance to be collected for viewing and analysis. Figure 6 shows examples of antenna gain plots for water-based and solvent-based inks on one of the packaging papers.

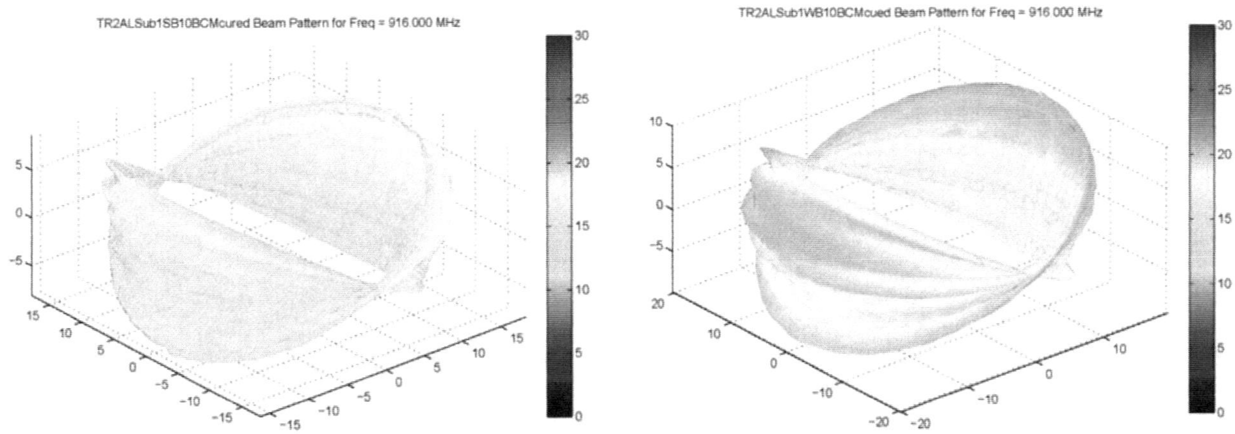

Figure 6 Antenna gain as a function of orientation for solvent based ink (left) and (water) based ink on a packaging paper.

Conclusions

Paper substrates are of great interest for printed electronics, because of extensive use in packaging. However, they pose a number of challenges for printing with functional materials. These challenges include surface irregularities and pores, nonuniform ink films and ragged lines. Suitable functional materials for printing ICs present additional challenges. We continue to progress at addressing these challenges and continue to identify and address new challenges.

Other results from printed antennae measurement are being published in various journals and have been or will be presented at appropriate conferences.

Acknowledgement

The authors would like to thank The Michigan Economic Development Corporation for full financial support for this work.

---

[1] Wood, L., Hrehorova, E., Joyce, T., Fleming, P. D., Joyce, M. K., Pekarovicova A. and Bliznyuk, V. "Paper Substrates for Printed Electronics", *Pira Ink on Paper Symposium*, Atlanta, September 28, 2005.

[2] Cruz, M., Joyce, M. K., Fleming, P. D., Rebros, M., "Surface Topography Contribution to RFID Tag Efficiency Related To Conductivity," TAPPI Coating and Graphic Arts Conference 2007, Miami, FL, April 21 – 25, 2007

[3] Stephen R. Forrest, "The path to ubiquitous and low-cost organic electronic appliances on plastic", Nature **428**, 911 (2004).

[4] MacDonald, W. A., "Engineered films for display technologies," Journal of Materials Chemistry, 14, 1 (2004) p. 4-10.

[5] D Chirvase, J Parisi, J CHummelen and V Dyakonov, "Influence of nanomorphology on the photovoltaic action of polymer–fullerene composites", Nanotechnology **15** (2004) 1317–1323.

[6] Hrehorova, E., Wood" L. K., Pekarovic, J., Pekarovicova, A., Fleming, P. D., and Bliznyuk, V., "The Properties of Conducting Polymers and Substrates for Printed Electronics", *Proceedings of IS&T Digital Fabrication 2005*, Baltimore, September 18-23, 2005, 197-202.

[7] Erika Hrehorova, Alexandra Pekarovicova and Paul D. Fleming, "Gravure Printability of Conductive Polymer Inks", *Proceedings of IS&T Digital Fabrication 2006*, Denver, September 18-23, 2006.

[8] Hrehorova, E., Rebros, M., Pekarovicova, A., Fleming, P. D., Bliznyuk, V. N., "Characterization of Conductive Polymer Inks based on PEDOT: PSS," Proc. of TAGA 59th Annual Technical Conference, Pittsburgh, PA, 18-21 March, 2007.

[9] Erika Hrehorova, Alexandra Pekarovicova, V.N. Bliznyuk, and Paul D. Fleming, "Polymeric Materials for Printed Electronics and Their Interactions with Paper Substrates", *Proceedings of IS&T Digital Fabrication 2007*, Anchorage, September 16-21, 2007.,

[10] A. Knobloch, A. Manuelli, A. Bernds and W. Clemens, "Fully printed integrated circuits from solution processable polymers", J. Appl. Phys. 96 (4) 15 August 2004.

[11] Adam Pron and Patrice Rannou, "Processible conjugated polymers: from organic semiconductors to organic metals and superconductors", Prog. Polym. Sci. **27**, 135-190 (2002).

[12] A. R. Brown, C. P. Jarrett, D. M. de Leeuw, M. Matters, “Field-effect transistors made from solution-processed organic semiconductors” Synthetic Metals **88** (1997) 37-55.
[13] H. E. Katz, A. J. Lovinger, J. Johnson, C. Kloc, T. Siegrist, W. Li, Y.-Y. Lin & A. Dodabalapur, “A soluble and air-stable organic semiconductor with high electronmobility”, Nature **404**, 478-81, 30 MARCH 2000.
[14] Blue Ramsey, David Harrison, Darren Lochun, Peter Evans and Paul Harrey, “Handbook of Conductive Lithographic Films”, http://www.brunel.ac.uk/research/cleaner/blue.html.
[15] Mark R. James, “Manufacturing Printed Circuit Boards Using Ink Jet Technology”, Proceedings of NIP19, International Conference on Digital Printing, Society of Imaging Science and Technology, New Orleans, Sept. 28-Oct. 3, 2003, p651.
[16] Toshihiko Oguchi, Keiki Suganami, Taizo Nanke and Toshihiko Kobayashi, “Formation of Precise Electrically Conductive Pattern Using Metal Colloid I-J Ink”, Proceedings of NIP19, International Conference on Digital Printing, Society of Imaging Science and Technology, New Orleans, Sept. 28-Oct. 3, 2003, p656.
[17] Rita Mohanty, “Silver Migration in Polymer Thick Film”, SGIA Journal, 8, 41-45 (2004).
[18] Woznicki, Tom, “Pads-Only Plating and Conductive Inks”, The Flex Circuit News [On Line], Available: http://www.flexdude.com/Back_Issues/FCN1-02A.PDF, 2002.
[19] Jennifer Rigney, “Materials and Processes for High Speed Printing for Electronic Components”, Proceedings of the IS&T NIP20: International Conference on Digital Printing Technologies, Salt Lake City, 2004, pp275-278.
[20] Groenendaal, L., Jonas, F., Freitag, D., Pielartzik, H., Reynolds, J. R. “Poly (3,4-ethylenedioxythiophene) and Its Derivatives: Past, Present, and Future,” Adv. Mater., 12, 7, (2000) p.481-494
[21] Kirchmeyer, S., Reuter K., “Scientific importance, properties and growing applications of poly(3, 4-ethylenedioxythiophene),” J. Mater. Chem., 15, (2005) p. 2077-2088
[22] Ashizawa, S., Horikawa, R., Okuzaki, H., “Effect of Solvent on Carrier Transport in Poly(3,4-ethylenedioxithiophene)/poly(4-styrenesulfonate),” Synthetic Metals, 153, (2005) p. 5–8
[23] Ouyang, J., et al., “On the mechanism of conductivity enhancement in poly (3,4-ethylenedioxythiophene):poly(styrene sulfonate) film through solvent treatment,” Polymer, 45, (2004) p. 8443–8450.
[24] Sedra, A. S., Smith, K. C., *“Microelectronic Circuits,”* 5th Ed., Oxford University Press, Oxford, NY, 2004 pp. 1283.
[25] Baude, P. F., et al., “Organic Semiconductor RFID Transponders,” *IEEE Int. Elect. Dev. Meeting Tech. Digest* (2003), p. 8.1.1-4.
[26] Kahn, B. E., “Developments in Printable Organic Transistors,” Pira International Ltd., Surrey, UK (2005), pp. 150.
[27] Fix, W., Ullman, A., Ficker, J., Clemens, W., “Fast polymer integrated circuits,” Applied Physics Letters, 81, 9 (2002) p. 1735-1737.
[28] Bartzsch, M., Fuegmann, U., Fischer, T., Hahn, U., Kempa, H., Preissler, K., Schmidt, G., Huebler, A., “All-printed electronics and its applications: a status report,“ Proc. on IS&T’s DF2006: International Conference on Digital Fabrication 2006, Denver, CO, September 18 – 22, 2006, p. 13-16.
[29] Gelinck, G., Geuns, T., Leeuw, D., “High Performance All-Printed Integrated Circuits,” Appl. Phys. Lett., 77, 10 (2000) p. 1487 – 1489.
[30] Andersson, P., et al., “Active matrix displays based on all-organic electrochemical smart pixels printed on paper,” Advanced Materials, 14, 20 (2002) p. 1460-1464.
[31] Kipman, Y., “Image quality Metrics for Printer s and Media”, IS&t’s 1998 PICS Conference, pp 183-187.
[32] S. Ramo, J.R. Whinnery, and T. Van Duzer, “Fields and Waves in Communication Electronics,” 2nd ed., John Wiley and Sons, New York, N.Y., 1984, Section 3.16 and 3.17, p. 147-154.
[33] “Electrical Resistivity of Pure Metals”, in CRC Handbook of Chemistry and Physics, Internet Version 2007, (87th Edition), <http:/www.hbcpnetbase.com>, David R. Lide, ed., Taylor and Francis, Boca Raton, FL, 2007.

607d Printed RFID Tags: Materials, Processes, and Devices

Vivek Subramanian, Daniel Soltman, Daniel Huang, and Steven Molesa. Department of Electrical Engineering and Computer Sciences, University of California, Berkeley, 571 Cory Hall, #1770, Berkeley, CA 94720-1770

Printed electronics has been an area of intensive research for the last several years. This has primarily been driven by a substantial reduction in price per unit area that is expected to result from the use of printing as a replacement technology for the lithography, etching, etc., used in conventional electronics fabrication. One of the most intriguing applications of printed electronics is as an enabling technology for the realization of ultra-low-cost RFID-based barcode replacement tags. By fabricating simple RF barcode circuits using printing, it is expected that it may be possible to realize such circuits at a net cost of <2 cents, thus enabling their deployment in such applications as anti-counterfeiting, authentication, and product inventory management.

To realize a printed electronic barcode, it is necessary to realize printed inductors, capacitors, transistors and diodes. This necessitates the development of printed conductors, dielectrics, and semiconductors, as well as their integration into functional devices with adequate performance. We will review our development of printable conductors based on nanoparticles, printable polymer dielectrics, and printable organic semiconductors. By optimization of materials and printing processes, we have successfully realized printed electronic devices with performance levels suitable for some basic RFID applications; here, we review the state of the art in our printed electronics technology and discuss the implications of the same on future smart tagging applications.

607e The Future of Smart Packaging and Rfid: The Path to Widespread Adoption (Panel Discussion)

Paul Rasband, Vue Technologies, 10894 Martingale, Frederick, MD 21701, **Jeff Lindsay**, Innovation Edge (www.innovationedge.com), 1526 South Commercial Street, Suite 200, Neenah, WI 54956, Gary F. Madsen, Auto-ID Sensing Technologies, Kimberly-Clark Corp., 2001 Marathon Ave., Neenah, WI 54956, Vivek Subramanian, Department of Electrical Engineering and Computer Sciences, University of California, Berkeley, Berkeley, CA 94720-1770, and Paul D. Fleming III, Department of Paper Engineering, Chemical Engineering and Imaging, Western Michigan University, Kalamazoo, MI 49008.

In spite of the many brilliant advances in RFID and smart packaging technology, there are challenges in identifying the proper business models to bring these technologies into widespread use. The five participants in this panel will share their insights into the future of these technologies and the pathway to adoption, and will answer questions from the audience.

MODELLING AND OPTIMIZATION OF WASHING PLANT OF A PAPER MILL

Amiya K. Ray,*Indian Institute of Technology Roorkee,India*

Abstract

In this present study a series of mathematical and statistical models are developed for various subsystems for optimizing a multi-stage brown stock washing plant interlinking with other operations and processes in a paper mill. Conjugate gradient (Fletcher-Reeves)-a method of constrained optimization is used. Because all interconnecting operations are discrete individually but continuous when it is integrated to maintain production of paper, the model predicted data are subjected to univariate /multivariate statistical linear or nonlinear regression models, to build up the link between two consecutive interacting operations. From the results of constrained nonlinear optimization of washing operation based on total annual cost, taking in to account all costs including BOD treatment cost, it is found that the decision variables, dilution factor (DF) and number of stages (N) are found of the order between 2.5-2.8 and number of stages of 4 respectively for minimum cost of washing.

Introduction

Modeling and simulation has become an indispensable tool now a days to assist in accruing resource saving in a complex, intricate and capital intensive industry like pulp and paper Industry.This has many mathematically complex subsystems.Brown stock washing(BSW) is a sub-system which is found to be interactive with multi stage bleaching(MSB), multiple effect evaporation(MEE),recovery boiler operation, and effluent treatment plant. Input and output parameters of each subsystem have an important impact on the operation of the entire pulp mill,hence on the economy of operation. Change in percentage solids to evaporator (feed concentration) is resulted due to the change in dilution factor, which will affect the steam consumption,steam economy and area requirement for subsequent evaporator operation.Similarly, dilution factor has an effect on solids going out with discharged pulp and its outlet consistency,and kappa number of the pulp going to the bleaching section.These parameters in turn will affect the chemical consumption and values of effluent parameters generated from bleach plant.Therefore, dilution factor (DF) and number of stages (N) of BSW have a visible impact on other subsystems.For a fixed set of output parameters for each subsystem an optimum DF and N are required to estimate.For this purpose, a functional relationship among output parameters, DF and N should be known. But no mathematical programming problem on multiple effect evaporator of black liquor coupled with brown stock washer and bleach plant is reported till the work under reference(1-4).The parameters for discrete subsystems are connected by statistical methods. By using regression analysis the functional relationships are thus obtained among the variables and are then subjected to mathematical programming for optimization of BSW. The objective of the problem is to minimize the inputs to the paper plant in terms of DF and N on various output parameters of the above operations and more importantly, to find appropriate models (statistical or mathematical) that could be used to predict the dependent variables (steam, power and chemical consumption, heat value of BLS etc.) at various of DF and N. Finally the optimized values of the parameters need to be evaluated through constrained nonlinear optimization techniques.The optimized results of DF and N need to be checked though classical graphical optimization which are still used in industry.

Development Of Models For Multi-Stage Brown Stock Washing

The objective function of multi-stage brown stock washer can be formulated as follows:
Total cost of washing (C_T) = Cost of washing Equipment (P_w) + Cost of maintenance and repair (P_m) + Cost of power due to drum rotation, back shower pump, vacuum pump, (P_o*C_{13}) + Cost of power for repulper-shredder (P_o*C_5)+ Cost of power due to liquor spray (P_o*C_6)+Cost of power for screw conveyer at the last stage of BSW (P_o*C_4) + Labour cost (C_{LW}) + Cost of steam for hot wash water ($P_A* C_A$) + Cost evaporation (P_E*C_E) + Cost of steam generated due to heat value of BLS ($P_E* C_L$) + Cost of condensate saving (P_s*C_s) + Cost of bleach chemicals due to carry over solids ($P_{NB}* C_{NB}$) + Cost of effluent treatment ($P_{Eff}*C_{Eff}$) + Cost due to soda loss ($C_{sl}* P_{sl}$) + Cost due to total solid loss ($C_{tsl} * P_{tsl}$)
Min (C_T) = $(P_w + P_m)*N + P_E* (C_E +C_A + C_L) + P_s* C_S + P_{NB}* C_{NB} + (C_{13} * N + C_5 * (N-1) + C_6 * N + C_4)* P_o + C_{LW} +P_{Eff}* C_{Eff} + C_{sl}* P_{sl} + C_{tsl}* P_{tsl}$ **(2.1)**
Where P and C being price/cost depending on the terms involved.C_{13} = power consumption due to drum rotation (C_1) + back shower pump (C_2) + vac.pump (C_3).

Multiple Effect Evaporator For Black Liquor Concentration

The modeling of multiple effect evaporators necessitates the knowledge of optimum number of effects based on prevailing costs. The mathematical models will be developed in terms of a system of non-linear algebraic equations depending upon how many number of evaporator bodies are there in the MEE.

Annual Cost Model for Optimum Number of Effects

For evaluating steam consumption and condensate saving, the required step is to develop an approximate annual cost of evaporation as a single valued function of number of effects (N) for which an objective function has been formulated as follows:
Total cost of evaporation (C_e) = Cost of labour (C_L) + Supervision charges (C_{SL}) + Cost of overall laboratory, administrative (C_{OLA}) + Fixed charges (C_{fc}) + Cost of cooling water (C_{cw}) + Cost of steam (C_s) + Periodic cleaning cost (C_c) + Cost of maintenance and repair (C_{mr}) + Cost of air handling equipment (C_{ae}) + Cost of equipment for condensing, injection (C_{ce}) + Cost of loss of capacity due to BPR (C_{BPR})
$C_e = \text{Min} (C_L + C_{SL} + C_{OLA} + C_{fc} + C_{cw} + C_s + C_c + C_{mr} + Ca_e + C_{ce} + C_{bpr})$ ... **(3.1)**
The Eq. 3.1 has been elaborated in detail **(1)**.Based on prevailing cost of equipments and utilities it has been found through both graphically and numerically that the optimum number of effects is six.

Models of Nth Effect Evaporator System

The mathematical model consists of four equations for each evaporators (i) total mass balance (ii) component mass balance (iii) enthalpy balance,heat transfer rate equations and (iv) equilibrium relationship for BPR. Therefore, sextuple effect evaporator system has 4N+1 i.e. 25 equations including overall mass balance equations. This has excluded the equations for heat transfer coefficient and physico thermal properties for black liquor with and without considering the condensate flash. Based on the approach of Holland (1), the design procedure of a sextuple effect evaporator system for truly backward feed sequence the mathematical model gives rise to a set of 12 non-linear algebraic equations in 12 unknowns and model with condensate flash gives rise to a set of 17 non-linear algebraic equations in 17 unknowns.

Case (I) Backward Feed (6→5→4→3→2→1)

$g_i = l_{n+1}\, T_s \{CP_{n+1} (u_{n+1} + BPR_{n+1}/T_s) - CP_n (u_n + BPR_n /T_s)\}/\lambda_s + (l_n - l_{n-1}) (\lambda_{n-1} + CPv\, BPR_{n-1})/\lambda_s - (l_{n+1} - l_n) \{(\lambda_n + CPvBPR_n) - CP_n\, T_s (u_n + BPR_n /T_s)\} /\lambda_s - (l_{n+1} - l_n) (\alpha\, u_n + \beta/T_s)\, T_s/\lambda_s$ .... **(3.2)**

$g_{i+1} = U_n\, T_s\, a_n \{u_{n-1} - (u_n + BPR_n/T_s)\} /50.\lambda_s - (l_n - l_{n-1}) (\lambda_{n-1} + CPvBPR_{n-1})/\lambda_s$ .... **(3.3)**

$(i=1,3......11;\ n=1,2,......6)$

Case (II) Backward Feed With Condensate Flash (6→5→4→3→2→1)

$g_i = l_{n+1}\, T_s \{CP_{n+1} (u_{n+1} + BPR_{n+1}/T_s) - CP_n (u_n + BPR_n /T_s)\}/\lambda_s + (l_n - l_{n-1} + m_{n+1}) (\lambda_{n-1} + CPv\, BPR_{n-1})/\lambda_s - (l_{n+1} - l_n) \{(\lambda_n + CPvBPR_n) - CP_n\, T_s (u_n + BPR_n /T_s)\} /\lambda_s - (l_{n+1} - l_n) (\alpha\, u_n + \beta/T_s)\, T_s/\lambda_s$ ... **(3.4)**

$g_{i+1} = U_n\, T_s\, a_n \{u_{n-1} - (u_n + BPR_n/T_s)\} /50.\lambda_s - (l_n - l_{n-1} + m_{n-1}) (\lambda_{n-1} + CPvBPR_{n-1})/\lambda_s$ ... **(3.5)**

$(i=1,3......11;\ n=1,2,......6)$

The general equation for 1st to 5th condensate flash tank

$g_i = \{(l_n - l_1 + m_1)\, \alpha\, T_s / \lambda_s)\} (u_{n-1} - u_n) - m_n\, T_s/ \lambda_s (Yu_n + ZT_s)$ **(3.6)**

$(i=13,14,....,17;\ n=1,2,...5)$

Modeling of Black Liquor properties (1-3)

Predictive equations for physico-thermal,physico-chemical and transport properties are required for modeling of multiple effect evaporator system. Regressions equations have been developed for all the relevant properties required for modeling and optimization of multiple effect evaporator systems. These properties are multi-variate in nature.

Solution of Models and Computer Programs

The Newton- Raphson-Jacobian matrix and Gauss Elimination are employed to solve a set of 12 non-linear simultaneous algebraic equations resulting from the multiple effect evaporator models. Computer program based on Fortran-77 has been developed to generate data on steam consumption, steam economy and area requirement for various values of dilution factor (DF), and number of stages (N). These are subjected to multivariate nonlinear regression equation for the BSW optimization. Total mass balance equations and component mass balance equations $L_i\, X_i$ are involved in the main program in terms of liquor flow rates.

Multi-Stage Bleach Plant –Modeling And Optimization

Development of Objective Function for optimization of bleach plant

An objective function to minimize the cost of bleach plant in terms of bleaching chemicals, and steam consumption has been formulated for a specific target variable (brightness in this case) and kappa number is as follows:

Total annual cost of bleaching (Z) = Base cost (C_o) + Cost of Chemical Consumption in stage 1 and 2 ($C_1 * x_1$) + Cost of chemical consumption in stage 3 ($C_2 * x_3$) + Cost of chemical consumption in stage 4 and 5 ($C_3 * x_5$) + Cost of steam in stage 1 and 2 ($C_4 * y_1$) + Cost of steam in stage 1 and 2 ($C_5 * y_3$) + Cost of steam in stage 1 and 2 ($C_6 * y_5$).

$Z = C_o + C_1\, x_1 + C_2\, x_3 + C_3\, x_5 + C_4 y_1 + C_5\, y_3 + C_6\, y_5$

Linear programming with constraints has been used to solve the problem with CEDED bleaching sequence as an example. The equations for estimating the neutralization cost due to consumption of NaOH has also been developed.

Development of Dynamic Model of a typical Multi-stage Bleach Plant

For multistage bleach plant like CEDED, one can develop a separate model for each stage but each stage does not have a large effect on the target variables (brightness, strength etc.). Therefore, it is prudent to assume one model to show combined effect of two stages. Using the above statistical method, the regression equations of Kappa Number (w_1), and residual chemicals (r), brightness (B) and brightness for nth stage (Bn), residual chemicals for nth stage (rn), can be written in nonlinear-orthogonal form, as follows:

$$w_1= w_{10}+a_1(w_2-w_{20}) +a_2(w_2-w_{20})^2+b_1(x-x_0)+b_2(x-x_0)^2+c_1(y-y_0)+c_2(y-y_0)^2+d_1(s-s_0)+d_2(s-s_0)^2 \quad \textbf{(4.1)}$$

$$r = r_0+a_3 (w_2-w_{20}) +a_4 (w_2-w_{20})^2+b_3(x-x_0) +b_4(x-x_0)^2+c_3(y-y_0) +c_4(y-y_0)^2+d_3(s-s_0) +d_4(s-s_0)^2 \quad \textbf{(4.2)}$$

$$B_n= B_{n0}+e_1\Delta w_{10}+e_2(\Delta w_{10})^2+f_1\Delta x_0+f_2(\Delta x_0)^2+g_1\Delta y_0+g_2(\Delta y_0)^2+h_1(\Delta s_0)+h_2(\Delta s_0)_2+j_1\Delta u_0+j_2(\Delta u_0)^2 \quad \textbf{(4.3)}$$

$$r_n= rn_0+e_3\Delta w_{10}+e_4(\Delta w_{10})^2+f_3\Delta x_0+f_4(\Delta x_0)^2+g_3\Delta y_0+g_4(\Delta y_0)^2+h_3(\Delta s_0)+h_4(\Delta s_0)^2+j_1\Delta u_0+j_2(\Delta u_0)^2 \quad \textbf{(4.4)}$$

Linearization of models

The nonlinear model equations are linearized by a general Taylor's series expansion and neglecting the higher order terms, the following equations yield:

$$w_1= w_{10} + a\Delta w_{20} + b\Delta x + c\, \Delta y + d\Delta s \quad \textbf{(4.5)}$$

$$r = r_0 +a\, \Delta w_{20} + b\Delta x + c\, \Delta y + d\Delta s \quad \textbf{(4.6)}$$

$$B_n = B_{n0} + e\Delta w_{10} + f\Delta x + g\, \Delta y + h\Delta s + j\Delta u \quad \textbf{(4.7)}$$

$$r_n = r_{n0} + e\Delta w_{10} + f\Delta x +g\, \Delta y + h\Delta s + j\Delta u \quad \textbf{(4.8)}$$

Where $\Delta x = x-x_0$

If a variable does not have a large effect, the variable would be dropped from the model for that particular stage. For each plant, there is a range of parameters in which the model is valid. For different set of ranges of parameters, one has to calculate new coefficients. Functional relationship for each variable has been developed by Multiple Regression analysis using the software named MATHCAD with the available experimental data.

Selection of Operating Constraints - the use of mathematical in-equations

It is well known that a model can be valid or accurate if it contains all the objectives, constraints, and decision variables relevant to the problem. The operating conditions are used as constraints because each plant has a limit for an operating variable (chemicals used, temperature, and retention time etc.) in a particular stage. These limits are put in the form of mathematical in-equations and can be obtained from plant study. These are then used as constraints for the LP.

Use of Linear Programming Technique and Computer Programs

The process of solving an LP requires a large number of calculations and is therefore best performed by a computer program. LINDO software is employed to quickly input an LP formulation, solve it, assess the correctness or appropriateness of the formulation based on the solution, and then quickly make minor modifications to the formulation and repeat the process. To solve the LP problem, LINDO requires operating data of various stages of CEDED bleach plant. The operational data presented by Freedmen are used to check their validity. It excellently matches without any noticeable error. For the above simulation cost operating data for the chemicals are estimated. Validity of the optimization scheme Models is made with the experimental data of Moore and Haner(1) and cost data reported by Freedman(4). The computed results agree very well without any error.

Calculation Procedure for Bleach Sequence

After model validation, one can now proceed for calculation for a specific case as follows:

To describe the optimization procedure for CEDED bleach plant, the derived models for first two stages can be written as:

$w_1 = a_0 + a_1\Delta w_{20} + a_2\Delta x_1 + a_3\Delta y_1 + a_4\Delta s_1$ **(4.9)**

$r = r_0 + a_5\Delta w_{20} + a_6\Delta x_1 + a_7\Delta y_1 + a_8\Delta s_1$ **(4.10)**

The model for third stage (chlorine dioxide) is as follows.

$B_3 = B_{30} + b_1\ \Delta w_1 + b_2\Delta x_3 + b_3\Delta y_3 + b_4\Delta s_3$ **(4.11)**

$r_3 = r_{30} + b_5\Delta w_1 + b_6\Delta x_3 + b_7\Delta y_3 + b_8\Delta s_3$ **(4.12)**

The model for fourth and fifth stage (extraction, chlorine dioxide) is

$B_5 = c_0 + c_1\ \Delta B_3 + c_2\Delta x_5 + c_3\Delta y_5 + c_4\Delta s_5$ **(4.13)**

$r_5 = r_{50} + c_5\Delta B_3 + c_6\Delta x_5 + c_7\Delta y_5 + c_8\Delta s_5$ **(4.14)**

It should be noted that the pH dependence is not included in this model as it is not very sensitive parameter and does not give accurate predictions through small changes in H+ ions. In general, the model can be expressed as:

$$\begin{pmatrix}\Delta B_1\\ \Delta B_2\\ \Delta B_3\end{pmatrix} = \begin{pmatrix} a_{11}\ a_{12}\ a_{13}\ a_{14}\ a_{15}\ a_{16}\\ a_{21}\ a_{22}\ a_{23}\ a_{24}\ a_{25}\ a_{26}\\ \end{pmatrix} \begin{pmatrix}\Delta X_1\\ \Delta X_2\\ \Delta X_3\\ \Delta X_4\\ \Delta X_5\end{pmatrix}$$

To make the calculation easy the equations for extracted kappa number from 1st stage (Eq.4.9), brightness from stage 3rd and 5th (Eqs. 4.11 and 4.13) of bleach plant model are used. Using the coefficient calculated the model Eqs. 4.9, 4.11 and 4.13 can be expressed as

$w_1=7.672+0.07*\ w_2-0.068*\ x_1-0.014*\ y_1-0.011*\ s_1$ **(4.15)**

$B_3=1.977-1.118*\ w_1+11.86*\ x_3+0.54*\ y_3+0.019*\ s_3$ **(4.16)**

$B_3=1.633+0.276*\ B_3+9.797*\ x_5+0.0.38*\ y_5+0.03296*\ s_5$ **(4.17)**

In stages IVth and Vth (D & ED stages) cost of relative amount of chemicals used is considered.

Using the coefficient for the conversion of stage temperature to steam costs and using prices for bleach chemical, the objective function can be expressed as:

$Z= -148.14 + 6.48*\ x_1+ 37.03*\ x_3 +17.89*\ y_1+1.08*\ y_3 + 2.16*\ y_5$ **(4.18)**

The objective function when subjected to the constraints (Eqs.4.15, 4.16 and 4.17) gives the optimum conditions. By using the above optimization scheme one can boost the significant economy by saving chemicals and energy in a bleach plant.

Development Of Statistical Models For Effluent Parameters

Development of models for estimating the pollution load generation in bleach plant is necessary as it will be extremely helpful in designing an effluent treatment plant (ETP). The extra cost due to carry-over solids and fluctuation of kappa number of pulp over a standard value can be ascertained. The cost can be expressed in terms of removal of any effluent parameters like BOD, COD, AOX etc. assuming proportional decrease of other parameters.

Development of Statistical Multi-variate Models of Pollutants in Bleach Plant to Estimate Pollutant Parameters

Multiple regression equations are also developed to estimate the effluent parameter (i.e. AOX, COD, BOD, TOCl, color etc.) and their inter-relationships from the chemical consumption due to either kappa number, carry-over solids or both from BSW. The total impact of effluent generation on BSW has been quantified using these inter-relationships and with BOD_5 or COD or AOX as a major variable which have been shown as a function of DF and N. These are also depicted in Table1.

Model Equations for Estimation AOX, TOCl, COD, Color, Polychlorinated Phenolics and Chloroform

There are three type model equations available in literature. The first one relates to all input variables and is empirical in nature; the second is a second order parabolic regression equation while the third is mostly used as a mathematically linear equations.

$$\mathbf{TOCl = k\ [ACl_2 + AClO^-/2 + AClO_2/5]} \quad \mathbf{(5.1)}$$

$$\mathbf{AOX = 1.2 - 2.0 * TOCl} \quad \mathbf{(5.2)}$$

Modification Of Existing Model

The regression equations are exercised for AOX, COD and COLOR. In this study, the statistical technique for single and multi-variate nonlinear regressions is adopted using **MATHCAD software.**

Statistical Regression Equation for AOX, COD, Color

Based on the data of **Basta et al (1)** the following multivariable nonlinear regression equation as a function of CF (Chlorine factor), %D (%Chlorine- di- oxide substitution) is developed.

$$AOX=- 0.218 +0.01 (\%D +2.065(CF)-8.714 E\text{–}05 (\%D)^2 -0.017 (CF* \% D)+8.0 E\text{-}03(CF)^2; R^2=1 \quad \mathbf{(5.3)}$$

Based on the data of **Tsai (1)** et al. a multivariable linear regression equation for COD developed as:

$$COD=-159.526\text{–}0.354 (ka)^2 \text{–}569.556 (KF)^2+3.345 (KF*ka)+17.408 (ka)+161.727(KF);R^2 = 0.976.. \quad \mathbf{(5.4)}$$

Using the data given by **Tsai (1) et al.** the following polynomial regression is as follows:

$$Color=283.745+0.732(ka)^2\text{–}1.75*103(KF)^2+7.875(KF*ka)-30.164(ka) +397.041(KF);R^2 = 0.907 \quad \mathbf{(5.5)}$$

$$Color = 45.296 + 0.027 (COD)^2 - 0.89 (COD); R^2 = 0.886 \quad \mathbf{(5.6)}$$

Development of Interrelationship between Various Pollution Parameters

Development of inter-relationship among various parameters is important to estimate one (e.g. BOD) from the other (e.g. COD).

Approximate Relationship between BOD and COD

There is no regression equation available in literature to estimate BOD. However, **Heimburger et al. (1)** presented a detailed account of experimental data of BOD and COD for various sequences and also for both softwood(SW) and hardwood (HW). In this present work the following regression equations for BOD are developed to get an order of estimate values:

Softwood (brightness 90%, Kraft unbleached kappa No. 30.6-32)

$BOD = - 2.597 + 0.32\ (COD);\ R^2 = 0.816$ **(5.7)**

Hardwood (brightness 90%, Kraft unbleached kappa No. 15.8-16)

$BOD = 0.313 + 0.172\ (COD);\ R^2 = 0.406$ **(5.8)**

Combined data of Hardwood and Softwood (linear and nonlinear)

$BOD = - 3.432 + 0.323\ (COD);\ R^2 = 0.862$ **(5.9)**

$BOD = - 0.275 + 0.184\ (COD) + 1.187 * 10^{-3}\ (COD)^2;\ R^2 = 0.871$ **(5.10)**

Relationship between Kappa Number and BOD (For CE stage)

BOD is linearly dependent on kappa number of the pulp.

Using the **experimental data (1)** for bamboo pulp, the following regression equations are also obtained:

(a)Linear Regression: $BOD_5 = 12.648 + 0.192\ (kappa);\ R^2 = 0.97$ ... **(5.11)**

(b) Polynomial Regression: $BOD_5 = 9.314 + 0.416\ (kappa) - 3.612 * 10\text{-}3\ (kappa)^2;\ R^2 = 0.983$...**(5.12)**

Relation between BOD and Soda loss

The data (1) shows that with the increase in washing loss, BOD increases. A relationship is obtained between BOD and washing loss as follows:

(a) Linear Regression: $BOD_5 = 3.167 + 0.386\ (SL);\ R^2 = 0.974$ **(5.13)**

(b) Polynomial Regression: $BOD_5 = 3.45 + 0.343(SL) + 1.214 * 10^{-3}(SL)^2; R^2 = 0.984$ **(5.14)**

Where SL = Soda loss

Cost Due To Effluent Treatment

Using the above equations cost due to effluent treatment

(An order of estimate figures) can be worked out based on published information on global conditions.

Models For Heating Value Of BLS

An algorithm is used to obtain the heating value of BLS for various DF and number of stages of washers. The relationship between steam generation due to heat value of BLS, DF and N is obtained by using multivariate nonlinear regression analysis.

Methodology For Statistical Analysis Of Data

Using the relevant algorithms and computer programs based on C++ language, the values of total solid loss, total soda and percentage solid to evaporator and the relevant correlations are obtained.

Table 1 Regression Equations

Parameters	Regression Equation and their Nature	R^2
Steam to evaporator	$3.66*10^{-3}(DF*N)+5.39*10^{-3}(N^2)-0.054(N)+2.43+0.073(DF)-2.207*10^{-3}(DF^2)$ Multivariate nonlinear regression	0.999
Steam to mill	$-16.92(DF*N)-31.407(N^2)+344.195(N)+6.093*10^3+183.132(DF)-14.481(DF^2)$ Multivariate nonlinear regression	0.96
Δ S	$-22.346(DF*N)-50.303(N^2)+520.644(N)+1.059*10^3+260.227(DF)-21.57(DF^2)$ Multivariate nonlinear regression	0.96
NaOH consumption	$5.426(DF*N)+8.486(N^2)-84.829*N+303.601-76.038(DF)+7.197(DF^2)$ Multivariate nonlinear regression	0.988
Soda loss	$4.25(DF*N (+6.593(N^2)-65-974*N+236.164\ 59.174(DF)+5.602(DF^2)$ Multivariate nonlinear regression	0.98
Hot wash water	$303.67+57.84(DF)-2.33*10^{-4}(DF^2)$ Single variable parabolic regression	1
Power consumption	$P_2=0.234+2.69*10^{-3}(DF)+3.78*10^{-5}(DF^2)$ Single variable parabolic regression	1
BOD	$2.635(DF*N)+3.222(N^2)-32.987*N+145.13-30.189(DF)+2.926(DF^2)$	0.975
AOX	$0.451(DF*N)+0.695(N^2)-6.969*N+29.937-6.29(DF)+0.596(DF^2)$	0.988

Interactive Optimization Of Brown Stock Washers

Conjugate Gradient (Fletcher-Reeves) method of constrained optimization- a nonlinear optimization technique is used to solve the present problem. Comparison of results of reveals that though there are significant contribution of each of the cost terms towards the total annual cost, some are less sensitive to the optimum values, namely DF and N. Probably this is the reasons, many previous workers overlooked inclusion of many cost contributing factors in their optimization scheme. Impact of AOX is not considered in the analytical optimization, as it is not the usual parameter in sharp contrast to BOD. It is customary to examine through graph as most of earlier results are explained through graphical construction.

Graphical Interpretation of data from Cost Equation for Optimizing Parameters

The overall annual cost equation; Eq.2.1 is subjected to various values of DF and N. All individual cost components as a functional dependence on DF and N are also estimated. Two cases have been considered for effluent treatment cost-one with AOX and the other with BOD_5. In this optimization scheme, cost benefit due to ΔS, the difference of steam production from a standard base is not considered, as it is evident from the analytical optimization that it has no impact on the optimized values. The optimized value from the mathematical optimization gives slightly higher value than those visually observed from graphical optimization. Neglecting BOD (or AOX) or neglecting steam generation benefits term yields the same result in graphical optimization technique. If both the above parameters are neglected the DF and N are found to be on the order of 2.5 and 4 respectively. It indicates that the optimum number of values of DF and N obtained from graphs are insensitive of the aforesaid parameters. These values are in close agreement with the values obtained from nonlinear constrained optimization considering all the costs.

Table 2 Comparison of results from Analytical and Graphical Optimization

Parameter	Graphical optimization		Analytical optimization	
	DF	N	DF	N
Including BOD	2.5	4	2.789	3.958
Δ steam production	-	-	2.5	4.19
Including AOX	4.5	4	-	-
Excluding effluent parameters	2.5	4	3.24	3.83
Excluding steam production condensate saving	3.0	4	3.66	3.77
Excluding steam production condensate saving and effluent parameter	2.5	4	3.67	3.79

Conclusion

Constrained nonlinear optimization of washing operation based on total annual cost taking in to account all cost terms including BOD treatment cost, it is found that the dilution factor (DF) and number of stages (N) are of the order of 2.789 and 3.958(≈4 as N can not be a fraction) respectively for minimum cost of washing. The optimized value from the mathematical optimization gives slightly higher value than those visually observed from graphical optimization. If BOD (or AOX) or benefits due to steam generation are neglected the DF and N are found to be on the order of 2.5 and 4 respectively. Therefore, the DF between 2.5-2.8 and number of stages of 4 can be taken as optimum values for the estimation of energy advantage and the extent of ecofriendlyness for brown stock washing..

References

1. Gupta Anjana (2002), Mathematical modeling and analysis problems for optimum operation, Ph.D. Thesis, Indian Institute of Technology Roorkee, Roorkee, India.
2. Gupta,Anjana,Ray,A.K.and Singh,V.P.(2000),"Optimization of multistage bleach plant",IPPTA Vol.13(1), March, pp.29-37
3. Ray A.K(2001) " Application of Mathematical and Statistical models to Pulp and Paper Industry- Proceedings of International Conference on Mathematical modeling",TATA McGraw Hill Company Ltd.January,pp.252-256.
4. Anjana Gupta, **A.K. Ray** & V.P. Singh(2001) –" Mathematical modeling & optimization of Multistage bleach plant"- Proceedings of International Conference on mathematical modeling", TATA McGraw Hill Company Ltd.January,pp.244-251

Modeling H Factor-Kappa Number For Kraft Pulping Of Lantana Camara Plant- An Experimental Investigation

Amiya K. Ray and Manish Kumar Puri,Department of Paper Technology, Indian Institute of Technology Roorkee,Roorkee,India,Paper Mill Road,Saharanpur,India

Abstract

The H – factor is used for process control purpose in Kraft cooking process which combines the effect of both temperature and time into a single variable representing the extent of the cook. It is the integral of the relative reaction rate with respect to time. It is an extremely important operating parameter in modern process control. For Lantana camara model the delignification kinetics and activation energy have not been established even today. Since for the estimation of H factor activation energy values are required, this needs delignification reaction to be performed through pulping experiments in laboratory. Therefore in the present work a reasonable number of kinetic experiments are carried out and activation energy of lantana camara plant is determined. Using the experimental evaluation of the value of activation energy, kinetic model for delignification is developed. Both Vroom's and Jagannadh et al. H factors were determined. Once the H-factor has been determined,further attempts have been made to relate kappa number and H factor through regression equations.

Introduction

H factor is extremely useful in mill operation. It enables changes to be made in the time-temperature cycle while maintaining a constant degree of delignification. Cooking cycles which give equal H – factors will produce pulps of equivalent yield and lignin content when other conditions are identical. The combination of time and temperature does not have any effect; as long as the H – factor is constant, the yield and lignin content will be the same. The greater the H – factor (whether induced by higher temperature, longer time, or both) the more delignification occurs. At the same H – factor (whatever the combination of time and temperature to reach it) the kappa number is constant. It is well known that for the calculation of H factor estimation of activation energy is important consideration.

Models Available For H-Factor Calculation

Vroom model [2]:

Vroom model provides a method expressing both cooking times and temperature in a single numerical variable called H factor, so that the time and temperature of any cycle can be expressed as a single numerical value. This can be changed to get properly cooked pulps. H- factor, represents the area under a curve drawn between relative reaction rate and time. The H – factor can be calculated as follows.

The extent, x, of any of the component reactions of the pulping process varies with time according to the kinetic equation

$$\left(\frac{dx}{dt}\right) = \text{k. f (composition)} \qquad (2.1)$$

Where k is reaction rate constant which depends on temperature. The above equation can be integrated to give

$$f(x) = \int k\,dt \qquad (2.2)$$

From Arrhenius' theory of chemical reaction, the specific reaction rate can be written as

$$k = Ae^{\frac{-E}{RT}} \qquad (2.3)$$

From the above equation the relative rate constant is developed as $k_r = (k/k_{100})$. Further Vroom assumed activation energy E of the order 32 kcal/mol or 134 kJ/mol. Then K_r can be written as (with R = 1.987 cal/g mol K)

$$k_r = \exp(43.2 - 16{,}113/T)$$

or

$$k_r = \exp(43.2).\exp(-16{,}113/T) \qquad (2.4)$$

From the above equation k can be written as

$$k = p.\, k_r \qquad (2.5)$$

Where p = A/ exp (43.2)

So, $$\int k\, dt = p \int k_r\, dt \qquad (2.6)$$

Vroom called the integral on the right hand side of the above equation as H – factor

$$H = \int k_r\, dt \qquad (2.7)$$

By rearranging above equations it can be written that

$$x = f(H) \qquad (2.8)$$

It follows that the extent of any of the component reactions for which the assumed activation energy of 32kcal/mol is appropriate is a unique function of H – factor, regardless of the nature of the cooking cycle. The H – factor is usually calculated by a numerical procedure that is equivalent to plotting the relative rate constant vs. cooking time and evaluating the area under the curve. Final H – factors values typically range from 1000 – 1500 depending on conditions for Kraft pulping. It can be used to estimate lignin content and yield**.** The curves of relative rate constant vs. temperature, temperature vs. time and relative rate constant vs. time can be used for understanding of the H factor.

H – Factor successfully predicts both lignin content and yield, but the same is not true for pulp viscosity. The H – factor is based on the assumption that delignification is a single reaction. While this model holds well, lignin is not actually a "pure" compound undergoing a single chemical reaction to achieve delignification. Rather lignin is a complex molecule which participates for many types of reactions occurring during the delignification process.

Model by Jagannadh et al. (1993)

Jagannadh et al [3] developed an analytical equation from first principles to predict accurately Vroom's H – factor during chemical pulping. The equation takes into account the variation in entering liquor temperature. The heating rate from the entering temperature to the cooking temperature is assumed to be linear. This assumption results in a nonlinear relative reaction rate. The use of the entering liquor temperature as an input variable in conjunction with the nonlinear reaction rate allows this equation to be used in modern cooking schemes like MCC and RDH.

Based on this model, the reaction temperature during pulping can be related to time as

$$T = T(t) \quad \text{when} \quad 0 < t < t_{heat} \qquad (2.14a)$$

$$T = T_{cook} \quad \text{when} \quad t_{heat} < t < t_{cook} \qquad (2.14b)$$

Where t_{heat} = heating time, h

T_{cook} = cooking time, h

T_{cook} = cooking temperature, K

H – Factor was then expressed as a combination of contributions from the heating period and the cooking period, as shown below:

$$H = \int_0^{t_{HEAT}} \exp\left(a - \frac{b}{T(t)}\right) dt + \int_{heat}^{t_{cook}} \exp\left(a - \frac{b}{T_{cook}}\right) dt \qquad (2.15)$$

For a linear rate of heating during the heating cycle, T (t) may be written as

$$T(t) = T_0 + t\left\{\left(T_{cook-} T_0\right) / t_{heat}\right\} \qquad (2.16)$$

Where T_0 is the initial, or entering liquor temperature. Substituting equation 2.16 into equation 2.15 and integrating over the temperature- vs. –time profile of the entire cook yields an analytical expression for the H – factor, as given by following equation

$$H = \left\{ t_{cook} + ZT^2{}_{cook}\left(1 - \frac{2T_{cook}}{b}\right)\right\}\exp\left(a - \frac{b}{T_{cook}}\right) - \left\{ZT^2{}_0\left(1 - \frac{2T_0}{b}\right)\right\}\exp\left(a - \frac{b}{T_0}\right) \quad (2.17)$$

Where $Z = t_{cook}\,[b\,(T_{cook} - T_0)]$. As shown above the initial liquor temperature may be used as an independent variable during the analytical determination of the H – factor.

The equation developed allows for different initial liquor temperature in the estimation of Vroom's H – factor. The analytical equation developed here has been designed to be useful in simulation and control schemes. The equation can also be used to estimate the H – factor for new pulping schemes like MCC and RDH, in which the entering temperature is a variable. The H- factor values estimated from the above equation has been claimed to agree closely with experimental values over a wide range of operation.

Experimental Work

Kappa number and lignin content were determined by the standard TAPPI test methods T 236cm-85 and T 222 om-88 respectively. For the determination of activation energy for the lantana camara plant, cooking was carried out at different temperatures between 150-170^{0}C with 20^{0}C interval. Lignin content was determined at various cooking time varying from 0 to 100 min with 20 min interval. The time required to attain the cooking temperature were 90 to100 min. The lignin content for various pulps obtained was determined as follows. The lignin content in raw material was determined to be 18.21%.

Table 1: Lignin content at different temperatures

Time in minute	Lignin Content of pulp % at 150^{0}C	Lignin Content of pulp % at 160^{0}C	Lignin Content of pulp % at 170^{0}C
0	17.98	17.31	16.8
20	15.03	13.8	14.2
40	13.2	11	10.6
60	12.4	9.1	8.623
80	12.12	8.9	6.45
100	-	8.7	5.08

Kinetic Studies

The kinetics of delignification can be expressed in the form of a rate equation of nth order as follows

$$-\frac{dl}{dt} = kl^n \quad (2.18)$$

This can be further expressed as

$$\ln\left(\frac{-dl}{dt}\right) = \ln k + n\ln l \quad (2.19)$$

According to equation 2.19 slope of the graph between ln(-dl/dt) and ln(l)will give the order of reaction (delignification). Therefore, for each temperature a graph was plotted to determine the order. Slopes from the graphs at different temperatures were around 1.27, 1.2 and 0.96(=1)for 150^{0}C, 160^{0}C,170^{0}C. The graphs are shown below.

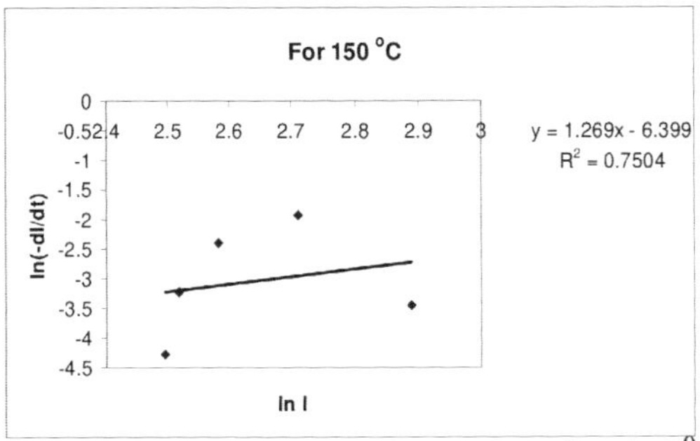

Fig 1: Graph for order of reaction for 150^{0}C

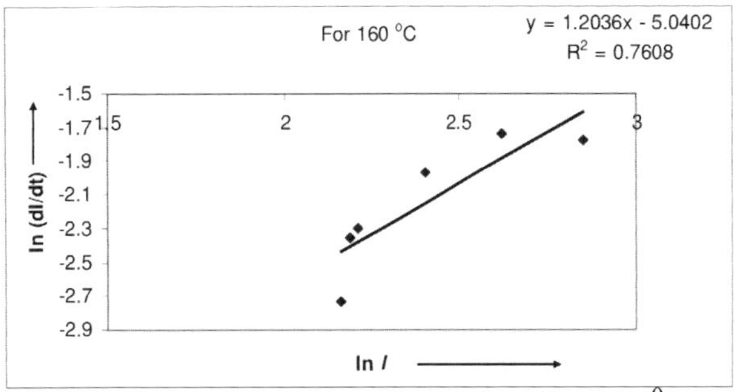

Fig 2: Graph for order of reaction for 160^{0}C

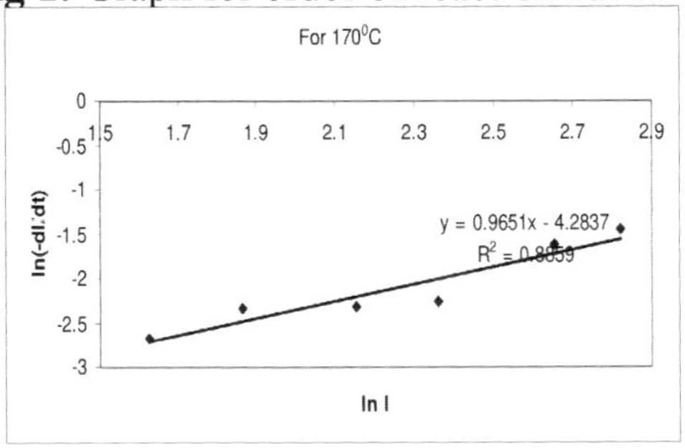

Fig 3: Graph for order of reaction for 170^{0}C

Based on the order of reaction relations were developed for the rate constant k in each case. The equations are as follows.

$$-\frac{dl}{dt} = kl^n$$

For order 1.27 (150^{0}C) ,the relation for k will be:

$$-\frac{1}{0.27}\left(\frac{1}{l^{0.27}} - \frac{1}{l_i^{0.27}}\right) = kt \qquad (2.20)$$

For Order 1.20(160^{0}C), the relation for k will be:

$$-\frac{1}{0.20}\left(\frac{1}{l^{0.20}} - \frac{1}{l_i^{0.20}}\right) = kt \qquad (2.21)$$

For order 1.00(170^{0}C), the relation for k will be:

$$\frac{1}{t}\ln\frac{l_i}{l_f} = k \qquad (2.22)$$

Finally the values of activation energy were calculated by the following equation.

$$-\ln\frac{k_1}{k_2} = \frac{E}{R}\left(\frac{1}{T_1} - \frac{1}{T_2}\right) \qquad (2.23)$$

Where, *l*i = Initial lignin concentration 18.21
*l*t= Final lignin concentration at time t
T_1, T_2 =two temperatures In degree Kelvin
R= Gas law constant =1.987 Cal/g mol K

Using the equations 2.20, 2.21, 2.22 and 2.23 the values of activation energy were calculated. The results obtained are tabulated below.

Table 2: Lignin Content for 1.5 Hours cooking:

Temperature	15%Sulphidity	17%Sulphidity	19%Sulphidity
150^0C	13.36	13.21	13.02
160^0C	9.94	9.88	9.63
170^0C	7.04	6.81	6.66

Rate constants are calculated from equations 2.20, 2.21, 2.22 calculated values are

Table 3: Values of Rate constant k for 1.5 Hours cooking:

Temperature	k at 15%Sulphidity	K at 17%Sulphidity	k at 19%Sulphidity
150^0C	0.039839493	0.041355819	0.043308191
160^0C	0.072039521	0.072804932	0.076055271
170^0C	0.633575149	0.655719182	0.670567606

Table 4: Values of activation energy E for 1.5 Hours cooking calculated using equation2.23;

Temperature	15%Sulphidity	17%Sulphidity	19%Sulphidity
$E_{150-160\,^0C}$	32571.56005	31593.98357	31504.49364
$E_{160-170\,^0C}$	37030.18861	37938.01979	37125.62589
Eavg	34800.87433	34766.00168	34315.05977

For the calculation of H factor grand average value of activation energy (E) was taken.as 34627.313 Cal/mole.

H- Factor Calculation

Using Vroom's model [2]

According to model developed by Vroom the H- factor, representing the area under a curve in which relative reaction rate is plotted against time.

$$\ln k = B - A/T$$

where A and B are constant for a given reaction .Constant A is equal to (E/R) where E is the activation energy of delignification reaction and R is the gas constant.

Therefore, the relation can be written as $$\ln k = 46.72 - \frac{17426.83}{T}$$

Based on these values of activation energy and rate constants from graph the frequency factors K_0 were calculated and the kinetic expressions were developed as follows:

For 150^0C: $$k = 1.29 \times 10^{15} \exp\left(\frac{34627.313}{RT}\right)$$

For 160^0C: $$k = 5.16 \times 10^{15} \exp\left(\frac{34627.313}{RT}\right)$$

Fig 4: Graphs for H factor

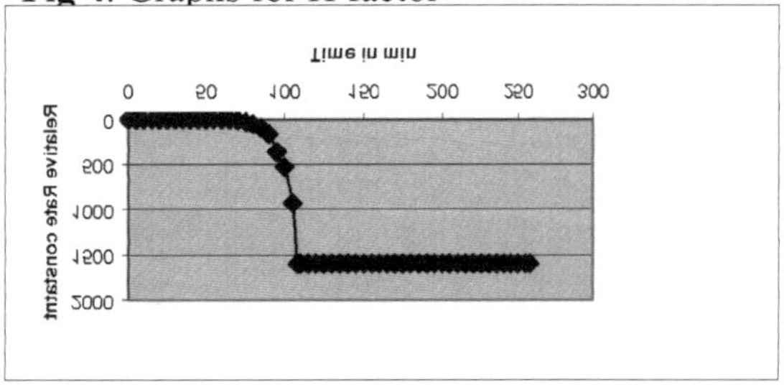

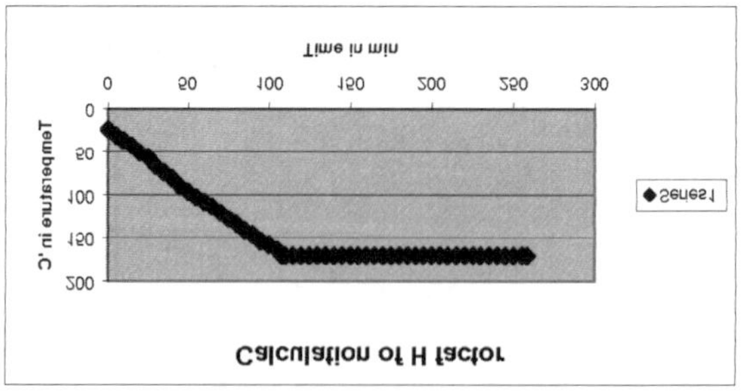

For 170^0C: $$k = 1.12 \times 10^{16} \exp\left(\frac{34627.313}{RT}\right)$$

Table 5: Values of Reaction rate constants

Temperature ^{0}C	k	Temperature ^{0}C	K	Temperature ^{0}C	k	Temperature ^{0}C	k
100	0.999	119	9.620	138	75.110	157	489.062
101	1.132	120	10.772	139	83.252	158	537.280
102	1.282	121	12.055	140	92.230	159	589.996
103	1.451	122	13.483	141	102.126	160	647.604
104	1.640	123	15.072	142	113.028	161	710.532
105	1.854	124	16.839	143	125.034	162	779.242
106	2.094	125	18.803	144	138.247	163	854.235
107	2.363	126	20.983	145	152.783	164	936.051
108	2.665	127	23.404	146	168.767	165	1025.276
109	3.004	128	26.090	147	186.335	166	1122.539
110	3.384	129	29.069	148	205.635	167	1228.523
111	3.810	130	32.370	149	226.829	168	1343.964
112	4.287	131	36.027	150	250.090	169	1469.655
113	4.820	132	40.076	151	275.610	170	1606.452
114	5.416	133	44.557	152	303.595	-	–
115	6.083	134	49.513	153	334.270		
116	6.827	135	54.992	154	367.878		
117	7.659	136	61.045	155	404.684		
118	8.586	137	67.730	156	444.975		

Table 6: Procedure for H factor calculation

Time,hr	Temperature^{0}C	Relative Reaction Rate Constant	Av. Rate	Time Interval	H
0	24	0	0	0	-
0.25	42	0	0	0	-
0.50	67	0	0	0	-
0.75	91	0	1.69	0.25	0.42
1.00	110	3.38	14.73	0.25	4.35
1.25	128	26.09	97.42	0.25	16.28
1.50	146	168.76	408.23	0.21	85.72
1.71	160	647.70	647.70		
1.75	160	647.70	647.70		
….	…..	…… .	……..		
….	…..	…….	……..		
3.21	160	647.70	647.70	1.5	971.55
				TOTAL=	1078.32

This is how H factor were calculated using the above procedure the values of H factor for pulpings carried out are shown below in the following table:

Table 7: Values of H factor using Vroom's model

Time, hrs	Tempe-Rapture, ^0C	H	Time, Hr	Temperature, ^0C	H	Time, Hr	Temperature, ^0C	H
1.5	150	417.47	2	150	541.99	2.5	150	660.66
1.5	150	427.54	2	150	552.23	2.5	150	666.51
1.5	150	424.4	2	150	541.99	2.5	150	666.51
1.5	**160**	**1078.91**	2	160	1417.91	2.5	160	1739.91
1.5	160	1103.25	2	160	1417.91	2.5	160	1598.4
1.5	160	1126.3	2	160	1398.25	2.5	160	1739.91
1.5	170	2649.34	2	170	3449.34	2.5	170	4249.34
1.5	170	2432.31	2	170	3356.87	2.5	170	4287.34
1.5	170	2649.34	2	170	3449.34	2.5	170	4249.34

Using model of Jagganadh et. al.[3]

$$H = \int_0^{t_{HEAT}} \exp\left(a - \frac{b}{T(t)}\right)dt + \int_{heat}^{t_{cook}} \exp\left(a - \frac{b}{T_{cook}}\right)dt$$

For a linear rate of heating during the heating cycle, T (t) may be written as

$$T(t) = T_0 + t\{(T_{cook-}T_0)/t_{heat}\}$$

Calculated values of H factor using model developed by Jaggandh are tabulated below;

Table 8: Values of H factor using model of Jaggandh et. al.

Time	**Tempe-rature**	**H**	**Time Hr**	**Temperature**	**H**	**Time Hr**	**Temperature**	**H**
1.5	150	443.26	2	150	608.16	2.5	150	708.15
1.5	150	468.96	2	150	618.52	2.5	150	718.52
1.5	150	473.98	2	150	608.16	2.5	150	718.52
1.5	160	1105.23	2	160	1497.91	2.5	160	1811.11
1.5	160	1198.80	2	160	1497.91	2.5	160	1788.32
1.5	160	1206.30	2	160	1477.66	2.5	160	1811.11
1.5	170	2712.01	2	170	3558.08	2.5	170	4346.69
1.5	170	2694.77	2	170	3410.35	2.5	170	4456.23
1.5	170	2712.01	2	170	3558.08	2.5	170	4346.69

Table 9: Comparison of Vroom H factor & H factor calculated based on model by Jagganadh et.al.

Time,hr	Temperature, ^0C	H Vrooms	H Jagganadh	Related error Based on Vroom
1.5	150	423.137	462.667	9.20%
1.5	160	1102.82	1170.11	6.12%
1.5	170	2576.99	2706.26	5.01%
2.0	150	545.40	611.613	12.14%
2.0	160	1411.35	1491.16	5.65%
2.0	170	3418.517	3508.33	2.62%
2.5	150	664.56	715.06	7.59%
2.5	160	1692.79	1803.51	6.54%
2.5	170	4262.06	4383.20	2.84%

Relation between H factor & Kappa number Kappa number and H factor values were plotted in figure 5 at various sulphidity conditions.

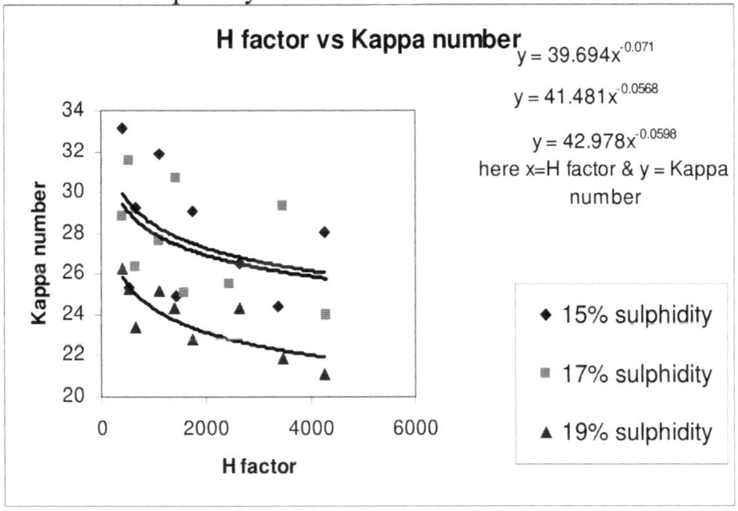

Fig 5: Relation between H factor and kappa number

Conclusion

From kinetics studies of Lantana camara plant following conclusions were made:

- Order of delignification reaction for Lantana camara plant was found 1.27, 1.2 and 0.96 for 150 ^{0}C, 160 ^{0}C and 170 ^{0}C.
- The activation energy of Lantana camara was found to be 34627.11 Cal/mol from kinetic studies.
- Models relating H factor and kappa number for pulpings found by experiments were
  - ✓ $Y=39.96X^{-0.071}$ (for 15% sulphidity)
  - ✓ $Y=41.481X^{-0.068}$(For 17% sulphidity)
  - ✓ $Y=42.987X^{-0.0598}$(For 19% sulphidity)
- Based on these values of activation energy and rate constants from graph the frequency factors K_0 were calculated and the kinetic expressions were developed as follows:

References

1) Kocurek, M.J., 1989, "Pulp and Paper Manufacture – Volume 5, Alkaline Pulping", TAPPI, Canada
2) Vroom, K.E., 1957, Pulp and Paper Magazine Can. 58:pp 228-31.
3) Jagannadh, S.V.N., Hart, P.W. and Hsieh, J.S., 1993, "An analytical equation for H-factor using the entering liquor temperature as a variable", TAPPI J 76(6):75
4) Octave Levenspile, 1999, "Chemical Reaction Engineering"3rd edition, John wiley and Sons ,New York

5) Biermann, C.J., 1996, “Handbook of Pulping and Papermaking”, Academic Press, CA.

647c Hot-Water Extraction Of Aspen Woodchips

Shijie Liu, Yang Wang, and Thomas E. Amidon. SUNY College of Environmental Science and Forestry, 1 Forestry Drive, Syracuse, NY 13210

This paper discusses the kinetic of hot-water extraction of Aspen woodchips. The Aspen woodchips are obtained from Alberta Pacific Forestry Industries, Inc., Boyle, Alberta. Besides the extractives, hemicelluloses are the most easily separable main component of woody biomass and thus form the bulk of the wood extracts obtained in the hot-water extracts. Xylan is the main polymeric moleculc in hardwood hemicelluloses. Xylan is a polymer made of β-xylopyranose units linked through (1→4)-glycosidic bonds, where arabinose, acetyl groups, and uronic acids are also present as lateral chains. Hydrolysis of hemicelluloses or xylan occurs simultaneously in the extract liquor during the extraction process. Hydrolysis produces 5-carbon sugars (mainly xylose), 6-carbon sugars (mainly glucose), and acetic acid.

647a New Pretreatment Method For Biofuels Production From Lignocellulosic Biomass

Yulin Deng and Yulin Zhao. School of chemical and bio-molecular Engineering, Georgia Institute of Technology, 500, 10th st. N.W., Atlanta, GA 30332

The technologies to convert lignocellulosic biomass into biofuels are rapidly evolving as a means to decrease dependence on fossil oil, reduce air pollution, and relieve the global warming phenomenon. A new pretreatment method was firstly proposed in this study for biofuels production from lignocellulosics. The concept of this pretreatment is to use a new developed carbohydrates dissolving method to slightly remove hemicelluloses and cellulose in lignocellulosic materials, disrupt the connections between hemicelluloses, cellulose and lignin, and alter the structure of treated biomass to make cellulose more accessible to the enzymes. We found that by partially removing hemicelluloses and cellulose, the wood chips were defibrated easily, and porous fibers and wood fiber particles were generated, which allows for hydrolytic enzymes to easily penetrate and hydrolyze the wood fibers, thus overcoming the key hurdle to the kinetics barrier of biological hydrolysis of lignocellulosics. In principal, the pretreatment method proposed in this study can be used for many kinds of lignocellulosic materials such as, wood, wastepaper, corn stover, rice straw, and switchgrass for biofuels production. It should be noted that this new pretreatment technology is effective for softwood species which the most current technologies could not hydrolyze efficiently. Therefore, the proposed technology provides a potential technique for biofuels production using abundant softwood species.

647e Increasing The Economic Efficiency Of Feedstock Used For Greenhouse Heating

Monty Singletary and Mark Bricka. Dave C. Swalm School of Chemical Engineering, Mississippi State University, Mississippi State, MS 39762

Greenhouses, while not a traditional forest product industry, can provide a stable outlet for waste biomass as a heat source. Originally using natural gas, the Natchez Trace Greenhouse in Kosciusko, MS, has recently converted to using alternative fuels for its heating needs. These greenhouses use biomass as a source of supplemental heat during the winter months, and the amount of land present at the site presents ample possibility for large quantities of biomass storage during the warmer months. Initially, corn was used as a fuel source, however, the cost of corn in Mississippi has risen almost parallel to that of natural gas. Waste sawdust from a local mill is available as a very cost effective fuel, but it alone does not provide adequate heating necessary for the number of greenhouses present on the property. This study aims to find an amendment that can be added to the waste wood to allow for ample heating, while meeting the greenhouse's monetary constraints. In addition, a small down-draft gasifier will be taken to the site to investigate gasification of this waste wood as an economically feasible alternative.

AUTHOR INDEX

AUTHOR INDEX